Studies in Systems, Decision and Control

Volume 593

The series "Studies in Systems, Decision and Control" (SSDC) covers both new developments and advances, as well as the state of the art, in the various areas of broadly perceived systems, decision making and control–quickly, up to date and with a high quality. The intent is to cover the theory, applications, and perspectives on the state of the art and future developments relevant to systems, decision making, control, complex processes and related areas, as embedded in the fields of engineering, computer science, physics, economics, social and life sciences, as well as the paradigms and methodologies behind them. The series contains monographs, textbooks, lecture notes and edited volumes in systems, decision making and control spanning the areas of Cyber-Physical Systems, Autonomous Systems, Sensor Networks, Control Systems, Energy Systems, Automotive Systems, Biological Systems, Vehicular Networking and Connected Vehicles, Aerospace Systems, Automation, Manufacturing, Smart Grids, Nonlinear Systems, Power Systems, Robotics, Social Systems, Economic Systems and other. Of particular value to both the contributors and the readership are the short publication timeframe and the world-wide distribution and exposure which enable both a wide and rapid dissemination of research output.

Indexed by SCOPUS, DBLP, WTI Frankfurt eG, zbMATH, SCImago.

All books published in the series are submitted for consideration in Web of Science.

Diogo R. M. Bastos · João Manuel R. S. Tavares

Advances in Gait-Based Identification

A Systematic Review of Deep Learning Models Leveraging Computer Vision Techniques

Springer

Diogo R. M. Bastos
Faculdade de Engenharia
Universidade do Porto
Porto, Portugal

João Manuel R. S. Tavares
Departamento de Engenharia Mecânica
Faculdade de Engenharia
Universidade do Porto
Porto, Portugal

ISSN 2198-4182 ISSN 2198-4190 (electronic)
Studies in Systems, Decision and Control
ISBN 978-3-031-89559-3 ISBN 978-3-031-89560-9 (eBook)
https://doi.org/10.1007/978-3-031-89560-9

This Springer imprint is published by the registered company Springer Nature Switzerland AG
The registered company address is: Gewerbestrasse 11, 6330 Cham, Switzerland

Diogo R. M. Bastos dedicates this book to his parents and to the journey that shaped him into a capable and determined individual. He is grateful for every gesture of love, support, and opportunity they have given to him. This work stands as a testament to the values and dedication that have inspired him.

João Manuel R. S. Tavares dedicates this book to his students, collaborators, colleagues and friends.

Foreword

Person identification has been a long dream since the 1800s when the Industrial Revolution paved the way for fast urbanization and mechanization. Growing cities started attracting people overnight looking for jobs, increasing the need for methods of people identification. Anthropometry, which involves measuring physical features such as height, weight, and the size of the head, is acknowledged to be one of the first formal methods to characterize an individual. Later on, scientists figured out that such information was not enough for people with matching body frames are likely to share similar data.

Fingerprints came further to provide a more reliable means of identification, followed by iris and face recognition. Despite being quite accurate, those methods require high-resolution sensors and the consent of a given individual. On the other hand, gait information can be acquired at a certain distance and with little knowledge of the person. Still understudied, gait-based biometrics is a fruitful area to complement and aid widely used methods for person identification, such as face recognition.

This book comes in handy and sheds light on an important research area, allied with an extensive analysis backed up by deep learning algorithms. The reader will be delighted to get a closer insight into the problem in the first two chapters, with datasets clarified and demystified further. If you are a deep learning addict, I am pretty sure you will be excited about the next chapter, where many techniques are evaluated and compared.

Last but not least, the last chapter provides a flavor of future directions and perspectives in gait-based person identification. After reading this book, if you have the "Black Mirror" syndrome, named after the famous TV show that explores the dark side of technology, you will change your pace every time you walk through a camera!

February 2025

Joao Paulo Papa
Universidade Estadual Paulista
Bauru, Brazil

Preface

The research area of this book, focused on person recognition through gait, has been attracting the attention of researchers, producers, and end-users.

Guided by the rigor that characterizes us, we embarked on a comprehensive systematic review that we consider thorough and transparent. This review provided a detailed overview of the current state of gait recognition, encompassing diverse datasets and existing methods for gait identification, categorized into deep-learning and non-deep-learning-based approaches.

Throughout this work, we discuss the substantial advantages of gait-based identification over traditional methods, and persistent challenges such as individual variations, clothing impact, and data quality issues. The convergence of different approaches was explored and expectations for improvements in the robustness and performance of gait recognition were also discussed.

We want to express our deep gratitude to Prof. João Paulo Papa for carefully reading the book and his willingness to write the foreword. Secondly, thank those who had an interest and dedication to exploring the knowledge contained in this book.

Finally, we hope this book can be a valuable resource for those interested in learning more about this promising and ever-evolving field.

Porto, Portugal

Diogo R. M. Bastos
João Manuel R. S. Tavares

Acknowledgements This work partially results from the project "Sensitive Industry", co-funded by the European Regional Development Fund (ERDF) through the Operational Programme for Competitiveness and Internationalization (COMPETE 2020) under the PORTUGAL 2020 Partnership Agreement.

Competing Interests The authors have no conflicts of interest to declare that are relevant to the content of this book.

Contents

Acronyms

AEI	Active Energy Image
ACL	Angle Center Loss
AI	Artificial Intelligence
ANN	Artificial Neural Network
BEI	Boundary Energy Image
BHCn	Batch Hard Contrastive Loss
BLSTM	Bidirectional Long-Short Term Memory
CapsNet	Capsule Network
CASIA	Institute of Automation, Chinese Academy of Sciences
CBAM	Convolutional Block Attention Model
CCGAN	Cycle-Consistent Generative Adversarial Network
CCGI	Color-mapped Contour Gait Image
CFFS	Composite Filter Feature Selection
CGI	Chrono-Gait Image
CJAM	Joint Attention Mechanism
CNN	Convolutional Neural Network
COCO	Common Objects in Context
CORF	Combination of Receptive Fields
CPD	Canonical Polyadic Decomposition
DAE	Deep Auto Encoder
DBN	Deep Belief Network
DCNN	Deep Convolutional Neural Network
DGEI	Dynamic Gait Energy Image
DINO	Data Independent Noise
DL	Deep Learning
DN	Discrimination Network
DoG	Difference of Gaussians
DTPC	Dilated Temporal Pyramid Convolution
DVS	Dynamic Vision Sensors
EER	Equal Error Rate
EGEI	Enhanced Gait Energy Image

ELM	Extreme Learning Machine
ff-GEI	Frame-by-frame Gait Energy Image
FPFE	Frame-level Part Feature Extractor
FQBlock	Frame Quality Block
GAEI	Generalized Active Energy Image
GaitST	Speed Transition Dataset
GAN	Generative Adversarial Network
GCEM	Gait Convolutional Energy Map
GCN	Graph Convolutional Network
GCS	Gait Cycle Segments
GEI	Gait Energy Image
GNN	Graph Neural Network
GQAN	Gait Quality Aware Network
HM	Horizontal Mapping
HPM	Horizontal Pyramid Mapping
IC-GEI	Incomplete Gait Energy Image
IGPS	Iterative Gait Period Set
IoT	Internet of Things
JIMEN	Joint Intensity Metric Estimation Net
KNN	k-Nearest Neighbors
LDA	Linear Discriminant Analysis
LGSD	Local Graphical Skeleton Descriptor
LSTM	Long Short Term Memory
MCM	Micro-motion Capture Module
MF-GEI	Multi-frequency Gait Energy Image
MLP	Multilayer Perceptron
MRL	Margin Ratio Loss
MSCAN	Multi-Scale Context-Aware Network
MSM	Mutual Subspace Method
NLPR	National Laboratory of Pattern Recognition
NMF	Negative Matrix Factorization
OF	Optical Flow
OU-ISIR	Osaka University—Institute of Scientific and Industrial Research
OULP	OU-ISIR Large Population
OUMVLP	Multi-View Large Population Dataset
PCA	Principal Component Analysis
PEI	Period Energy Image
PFI	Part-Frame-Importance
PQBlock	Part Quality Block
PSN	Pairwise Similarity Network
RBF	Radial Basis Function
RC-GEI	Complete Gait Energy Image
ResNets	Residual Networks
RNN	Recurrent Neural Network
ROI	Regions-of-Interest

RPN	Region Proposal Network
SDTW	Soft Dynamic Time Warping
SG	Smart Gait
SGCN	Spatial Graph Convolutional Network
SNN	Siamese Neural Network
SSM	Silhouette Stereo Map
ST-GCN	Spatial-Temporal Graph Convolutional Network
SVM	Support Vector Machine
TCN	Temporal Convolutional Network
TCNN	Two-Branch Convolutional Neural Network
TS-GAN	Two-Stream Generative Adversarial Network
VTM	View Transformation Model

List of Figures

List of Tables

Chapter 1
Introduction

Abstract This chapter presents the problem of identifying people through gait, a promising biometric method with significant benefits over traditional methods. However, challenges such as variations in gait due to different factors and the complexities of the environment still need to be overcome. Objectives are also presented, such as conducting a systematic review of deep learning models for gait-based identification, analyzing and comparing existing methods, and investigating current challenges. Finally the end, the book's structure is presented, ranging from fundamental concepts to method comparisons and conclusions.

1.1 Brief Description of the Problem

Person recognition through gait is one of the most promising biometric techniques, with tremendous advantages over traditional methods. Among other advantages, this technique is non-invasive, the data is easy to collect at a distance and can be used in uncontrolled scenarios [1].

Since 2014, Deep Learning (DL) based approaches have shown promising results in gait recognition by automatically extracting features [2]. However, there are still some issues that need to be addressed. Covariant factors such as changes in viewing angles, variation in walking speed, types of footwear, and the carrying of several objects or wearing different clothes can influence changing specific gait characteristics [1, 3]. The complexity and variability of environments and different ways of human body representations can also constitute a challenge for gait recognition [2].

Developing robust gait recognition systems capable of overcoming these challenges and limitations is essential to successfully implement them in real-world applications [2].

D. R. M. Bastos and J. M. R. S. Tavares, *Advances in Gait-Based Identification*, Studies in Systems, Decision and Control 593,
https://doi.org/10.1007/978-3-031-89560-9_1

1.2 Objectives and Results to Be Achieved

The primary objective of this book is to conduct a systematic review of deep learning models leveraging computer vision techniques for gait-based identification. The specific aims include:

- To explore the advancements in gait-based identification techniques, particularly focusing on deep learning models;
- To analyze and compare existing or originally proposed computational methods for gait-based human recognition and classification from digital images;
- To investigate the challenges and limitations of current computational gait identification systems, especially in addressing covariant factors such as changes in viewing angles, variations in walking speed, and environmental variables;
- To assess the feasibility and effectiveness of deep learning approaches in overcoming these challenges and enhancing gait-based identification systems.

The book aims to provide a comprehensive understanding of the state of the art in gait-based identification, highlighting innovative solutions and areas for further research. By synthesizing existing knowledge and advancements in the field, this review seeks to contribute to the ongoing development of robust gait recognition systems with applications in security, monitoring, and access control.

1.3 Book Structure

The structure of the subsequent chapters and sections in this work unfolds as follows: First, Chap. 2, a background chapter, provides fundamental insights into concepts pertinent to the systematic review undertaken. This is followed by Chap. 3, which explores research questions and data extraction methodologies using the PRISMA diagram. Also all the exclusion criteria applied to reach the final set of articles for the review were here highlighted. Chapter 4 provides an analysis of various datasets used in gait identification, comparing the accuracy values of different gait identification algorithms. A categorization and comparison of the different reviewed methods is made in Chap. 5. This includes traditional gait recognition approaches and deep learning-based model-free and model-based methodologies. Finally, the book culminates in a conclusions chapter, Chap. 6, synthesizing key findings and insights gleaned from the comprehensive review.

References

1. Harris, E., Khoo, I.-H., Demircan, E.: A survey of human gait-based artificial intelligence applications. Front. Robot. A **I**, 8 (2022). https://doi.org/10.3389/frobt.2021.749274
2. Khaliluzzaman, M., Uddin, A., Deb, K., Hasan, M.J.: Person recognition based on deep gait: a survey. Sensors **23**(10) (2023). https://doi.org/10.3390/s23104875
3. Russel, N.S., Selvaraj, A.: Gender discrimination, age group classification and carried object recognition from gait energy image using fusion of parallel convolutional neural network. IET Image Process. **15**(1), 239–251 (2021). https://ietresearch.onlinelibrary.wiley.com/doi/pdf/10.1049/ipr2.12024, https://doi.org/10.1049/ipr2.12024

Chapter 2
Background

Abstract This chapter is a fundamental introduction to the central role of gait recognition in biometrics, particularly in the context of Artificial Intelligence (AI). It elucidates the importance of gait as a behavioral biometric modality, offering advantages such as contactless and unobtrusive data collection. Focusing on vision-based gait recognition using image sensors, the chapter delves further into the multifaceted applications of gait recognition, covering person identification, re-identification, and authentication. Challenges, referred to as covariant factors, such as viewing angles, walking speed, and environmental variables, are discussed, highlighting the adaptability required for robust gait recognition systems. Finally, the chapter introduces the concept of gait feature representation, contrasting model-based and model-free approaches, laying the groundwork for subsequent discussions on DL methods for gait recognition. The typical Convolutional Neural Network (CNN) architecture is presented as a stepping stone for a nuanced exploration of other architectures. The concepts covered lay the groundwork for advanced investigations in subsequent chapters, contributing to the construction of necessary knowledge in the field of gait recognition.

2.1 Gait as Biometric Modality

One of the main applications of Artificial Intelligence (AI) techniques in studies of human gait is biometry [6]. Biometry is a technology to recognize the claimed identity [10]. A gait is a behavioral biometric that, through how people walk, allows the discovery of important information about them [8].

Smart Gait (SG) is the term used to refer to any integrated system for analyzing human gait data using AI. SG systems can analyze one or several subjects simultaneously [6].

Person recognition through gait is an active research area and one of the most promising biometric techniques, presenting tremendous advantages compared to traditional methods like face, iris, and fingerprint recognition [6]. One notable strength of gait recognition is its non-contact and inconspicuous data collection process. Unlike conventional biometrics requiring physical contact, gait recognition enables

D. R. M. Bastos and J. M. R. S. Tavares, *Advances in Gait-Based Identification*, Studies in Systems, Decision and Control 593,
https://doi.org/10.1007/978-3-031-89560-9_2

the unobtrusive acquisition of walking patterns from a distance, ensuring subjects remain unaware. Moreover, its non-invasive nature makes it effective even at lower resolutions, where other biometric methods might falter due to the lack of requisite high resolution [7].

In essence, gait recognition is a valuable biometric technique, showcasing advantages in discreet data collection and reliable operation at varying resolutions. These attributes render it well-suited for healthcare monitoring, security systems, and indoor and outdoor surveillance applications, where maintaining a non-intrusive approach is paramount. The continued exploration of gait recognition promises advancements in biometric technology with implications for diverse fields [7].

Over the past few decades, rapid and significant progress has been observed in gait-based recognition. This progress has been driven, in large part, by advances in sensing technologies, automation, data analysis, parallel processing, and the Internet of Things (IoT) [6]. Since 2014, the main contribution to progress has been developments in DL, showing promising results in gait recognition by automatically extracting features [8].

There are different sensing modalities, which can be classified into wearable or non-wearable. Wearable sensors are used to obtain the gait data from gait sequences. The non-wearable systems of interest for this work use imaging sensors, e.g., imaging cameras, to acquire gait information, a process called vision-based gait recognition [8].

2.1.1 Applications

Gait recognition is a term used to refer to all applications that use gait as a soft biometric feature to identify people based on their way of walking. These main applications include person identification and re-identification, person authentication, gender recognition, age estimation, crowd monitoring, and multiple player tracking and identification [6]. Of the applications mentioned, the most complicated to distinguish are gait identification, re-identification, and person authentication.

Identification

The process of identification in gait recognition involves associating the observed gait patterns with the unique characteristics of a previously registered individual. Gait is a distinctive biometric feature that serves as the key identifier, allowing the system to discern and attribute the correct identity. This application is particularly useful in scenarios where the primary objective is to recognize individuals at different points in time and involves a one-to-many verification [15]. Usually, the computational system achieves this by mapping the current gait observations to the corresponding identity within its database of registered subjects. The identification system is trained to capture and differentiate the nuances in each individual's walking pattern, making it robust to variations caused by clothing, carrying conditions, and walking speed. By focusing on the intrinsic gait features, the identification process ensures reliable and accurate recognition across different instances [17].

Re-identification

Re-identification in gait recognition expands the scope beyond singular instances of identification, aiming to recognize and track individuals across diverse imaging cameras or locations over time. This application is particularly relevant in multi-camera surveillance environments, where the system facilitates continuous tracking of people as they move through different areas [5].

One of the key challenges addressed by re-identification systems is the seamless transition between cameras, which may have varying viewpoints and environmental conditions. These systems incorporate advanced algorithms to overcome obstacles like temporary occlusions and lighting variations, ensuring a consistent association between the observed gait patterns [5].

In addition to gait features, re-identification systems may integrate other biometric and contextual characteristics to enhance accuracy. This comprehensive approach allows the system to adapt to changing scenarios and maintain a continuous tracking mechanism, even when individuals are acquired by different cameras under diverse conditions. The goal is to establish and preserve identity links across various instances and locations, providing valuable insights into surveillance and security applications [16].

Authentication

The emphasis of gait recognition can be on authentication. Unlike gait identification, where the result of the algorithm provides a direct classification indicating the person's identity, in the context of gait authentication, the result is not an explicit classification. Instead, the algorithm generates a score representing how similar the presenting person's gait is to the previously recorded gait. This score is a measure of similarity, indicating the degree of correspondence between the gait characteristics of the person in question and the stored references. The higher the score, the greater the similarity, suggesting a closer match between the gaits. In this case, the network would be trained to authenticate a person based on the similarity or difference between the gait presented and the stored references [7].

Employees would initially be registered in the system by sampling their movements. This could be done using existing surveillance cameras or dedicated cameras. The system would be deployed in restricted areas, where cameras continuously acquired the gait of people approaching. When a person tries to access a restricted area, the system compares the current gait with the stored samples in a one-to-one verification [15]. Access is granted if the gait is recognized as belonging to the authorized employee. Otherwise, access is denied.

Gait is a unique biometric characteristic that is difficult to replicate, which increases security compared to conventional methods such as access cards or passwords.

2.1.2 Covariant Factors

The successful computational implementation of gait recognition is intricately tied to its ability to adapt to many challenging scenarios, each influenced by a spectrum of covariant factors that define its application context. As one delves into these challenges, it becomes evident that mitigating their impact is crucial for the overall efficacy of gait recognition systems.

A pivotal challenge arises from the diverse variations in viewing angles, a factor magnified by the limitations of cameras acquiring images from singular perspectives. This limitation becomes particularly pronounced in cross-view scenarios, where images of the same subject appear markedly different. Navigating this intricacy is paramount for feature extraction processes to yield robust and reliable outcomes in gait recognition [18].

In addition to viewing angles, variations in walking speed introduce one more layer of complexity, influencing specific gait characteristics. The system must adapt to these variations, ensuring accurate recognition irrespective of the pace at which an individual moves. Footwear choices further contribute to the challenge, as different types of shoes can alter the overall gait pattern. Likewise, individuals carrying objects or donning diverse clothing attire present additional variables that demand consideration in the gait recognition process [6, 13].

The environmental context, marked by its inherent complexity and variability, introduces another dimension of challenge. Gait recognition systems must grapple with the unpredictability of real-world settings, accommodating diverse surroundings and acquiring human movement patterns under different circumstances. Moreover, the various ways the human body can be represented, influenced by factors such as attire and posture, further underscore the intricate nature of gait recognition [8].

In essence, addressing these covariant factors requires a nuanced approach, where computational gait recognition systems must exhibit adaptability, resilience, and a capacity to discern patterns amidst myriad influencing elements. The journey towards robust gait recognition involves technological advancements and a deep understanding of the contextual intricacies that shape the field.

2.2 Gait Feature Representation

Gait feature representation refers to how the human body is encoded or modeled in a gait recognition system. This representation is fundamental to computational gait recognition methods, as it describes how information related to human anatomy is acquired and used to identify or authenticate individuals based on their gait. Gait feature representation is intrinsically linked to gait recognition methodologies, especially when discussing the choice between model-based and model-free approaches [7]. The choice of how to represent the gait will depend on which of the two methodologies to use.

2.2.1 *Model-Based Approaches*

Model-based gait approaches can work with kinetic parameters or spatiotemporal parameters [4].

Spatiotemporal parameters include features such as gait cycle period, phase length, and walking speed [4]. The model-based approaches with kinetic parameters involve explicitly constructing a three-dimensional or two-dimensional human anatomy model, including specific features such as body segments and joints. This type of methodology is more complex than model-free approaches but more accurate and resistant to problems of cross-view [7].

To achieve robustness, the model-based methods use a dynamic approach, which means they not only acquire a static image of the human anatomy but also monitor and record the changes in these structures over time. This involves tracking how body parts move and change as a person walks. This dynamic monitoring helps overcome noise and occlusions [8].

In this context, gait feature representation refers to the geometric modeling of anatomy, where the body's structure is described in terms of three-dimensional (3D) or two-dimensional (2D) entities. This can include representations such as skeletons, anatomical landmarks, or even detailed 3D models.

2.2.2 *Model-Free Approaches*

The distinction between model-based and model-free approaches is often related to the complexity of the used model and how the information is represented. In this case, a mode-free methodology, also called appearance-based [6], looks at the human body's overall movement based on the person's complete silhouette. By using silhouettes, the gait recognition system does not focus on specific anatomical details, such as skeletons or characteristics of individual body parts. Instead, it considers the body's overall shape, acquiring the silhouette while walking. This choice is often made for its computational efficiency and ability to handle low-resolution images [7, 8].

A computational gait recognition system based on a model-free approach normally starts with a human detection and tracking module. This module extracts silhouettes in different frames from an input video, reconstructing and subtracting the background from each image. This process allows the extraction of the object in

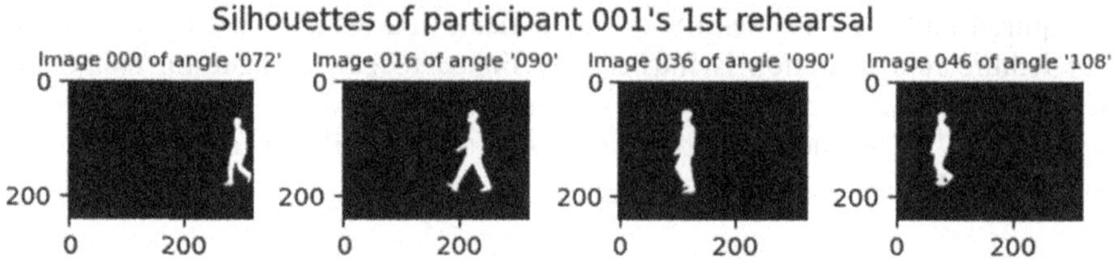

Fig. 2.1 Examples of silhouettes in three angle variations in data for a specific person

motion. It is then necessary to apply post-processing filters to remove noise, which is essential to generate images independent of textures such as clothing colors [7].

The result is white silhouettes on a black background, as shown in Fig. 2.1. These silhouette images were selected from the popular gait benchmark dataset CASIA-B, which is covered in more detail in Chap. 4.

In Fig. 2.1, each image represents a frame from the input video and can be used directly in certain methodologies adopted for gait recognition. However, some appearance-based gait recognition methods use more compact and informative representations of a person's gait, computing average features over a complete gait cycle or segregating the gait cycle into key poses. Key pose-based approaches divide the gait cycle into meaningful key poses. This allows you to preserve gait dynamics in high resolution [4].

Gait Energy Image

Gait Energy Images (GEIs) are widely used model-free human recognition algorithms once, with a simple representation of gait data, the noise can be removed being relevant gait information preserved [6].

A GEI is the result of averaging the silhouettes of a gait cycle. This integrates static and dynamic components of human gait, including frequency of movement and temporal and spatial changes that influence the appearance of the human body [13].

Both spatial and temporal information is then represented in a single grayscale image, which considerably reduces the computational effort compared to using a sequence of silhouettes [6, 14]. This time-normalized image has pixels with different intensities. The pixel with the highest intensity in the GEI indicates that human movement occurred most frequently at that position [2]. As examples of GEIs, one can consider those present in Fig. 2.2 obtained using the silhouettes on the CASIA-B dataset.

There have been efforts to replace GEI with representations that preserve more temporal and spatial information. However, this is still widely used.

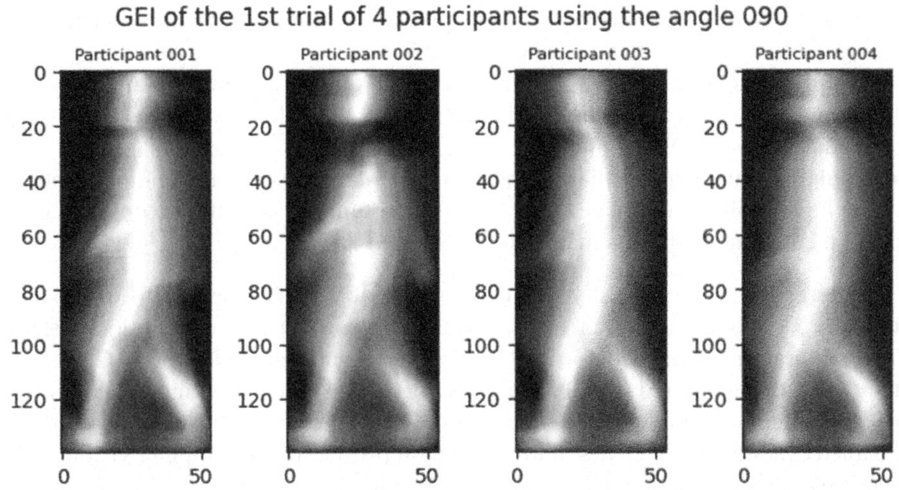

Fig. 2.2 Four examples of GEIs from four participants in the CASIA-B dataset using the angle 090

2.3 Deep Learning Methods for Gait Recognition

In the process of gait recognition, there are different DL-based methods with different deep architectures. The choice of the appropriate method may depend on the specific needs and characteristics of the application in question. Examples of factors that can influence the choice of method are:

- Use environment: The environment where the gait recognition system will be applied is a critical factor influencing the control of covariant factors. In outdoor environments, such as surveillance systems in public spaces, it may be necessary to deal with greater variability of conditions. Lighting, for example, can vary significantly, influencing the choice of image processing and analysis techniques. On the other hand, in controlled environments, such as indoor spaces with entry control systems that require people to pass through one at a time, it is possible to use more precise and sophisticated computational methods;
- Number of Cameras: If the application involves multiple cameras for data acquisition, then the integration and synchronization of information from multiple sources may be necessary. This relates to an operation known as "tag-and-track", which includes identifying a subject from one camera and matching it with people in other cameras with non-overlapping fields of view [6]. People re-identification applications are widely used in security systems and video surveillance. This can influence the choice of data fusion and object-tracking computational methods;
- Recognition Objectives: If the application aims to identify people in a known database, the analysis can focus on comparison and matching techniques based

on specific gait characteristics. If the application is for authentication, verification and validation techniques may be preferable;

- Real-Time Requirements: In applications that require real-time recognition, methods must be efficient and fast. This can influence the choice of image processing and analysis and DL techniques;
- Available Data: Data availability is also important. In some applications, obtaining high-quality gait images may be difficult, limiting algorithmic options. Date augmentation algorithms may also be required.

Therefore, given the diversity of methods available, it is essential to understand how each of them can adapt to the specific needs of a gait recognition system. The exploration is primarily focused on Convolutional Neural Networks (CNNs), as they constitute the foundational understanding crucial for delving into the intricacies of other deep neural network architectures. The forthcoming subsection provides a detailed examination of CNNs, unraveling their pivotal role in computer vision. Subsequently, this foundational knowledge will serve as a stepping stone for a nuanced exploration of other architectures, each contributing to the multifaceted landscape of deep learning in gait recognition.

2.3.1 Convolutional Neural Network

A Convolutional Neural Network (CNN) is a specialized variation of Neural Networks that has shown exemplary performance in several competitions related to computational vision, image processing and analysis, and classification [9].

There are several variants of CNN architectures in the literature. However, their basic components are very similar. Taking the well-known LeNet-5 as an example, it is composed of three types of layers: convolutional, pooling, and fully connected layers, as shown in Fig. 2.3 [3].

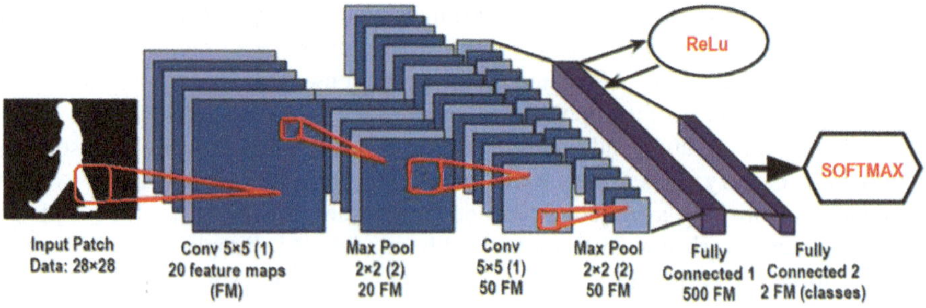

Fig. 2.3 The general architecture of LeNet. The numbers in the convolution and pooling layers indicate the kernel size, stride (in parentheses), and the total number of neurons present in each layer (adapted from Yap et al. [19])

Convolutional Layer

A convolutional layer is key in extracting features from an image or any other input with a similar structure. This extraction occurs by applying convolutional filters to the input, which produces feature maps as output, as shown in Fig. 2.4. These filters are small arrays of weights that slide across the input to perform the convolution operation [9, 11, 12].

The filter is applied to different input regions during convolution through an elementary multiplication and sum operation. This operation calculates the dot product between the input and filter values at each position, resulting in a unique value in the output feature map. This process is repeated at all positions in the input, producing a new feature map [1, 3].

The convolutional layer can also include other important steps, such as applying an activation function, e.g., ReLu or Sigmoid, on the feature map values to introduce non-linearity [12].

The convolutional layers in a CNN are designed to automatically learn hierarchical and locally invariant features in the inputs, such as edges, textures, and patterns relevant to the task at hand. These layers allow CNN to process images effectively and other similarly structured inputs, making it a powerful choice for computational vision tasks such as image classification, object detection, and semantic segmentation [3, 9].

Pooling

In addition to storing all the distinctive features of an image, convolutional layers can reduce the amount of data to be processed [1]. However, it is pooling techniques that have the true ability to reduce dimensionality [9]. This reduces the number of parameters to be learned by the model, which reduces computational effort and avoids overfitting problems. There are several types of pooling; however, the most used are max pooling and average pooling [1].

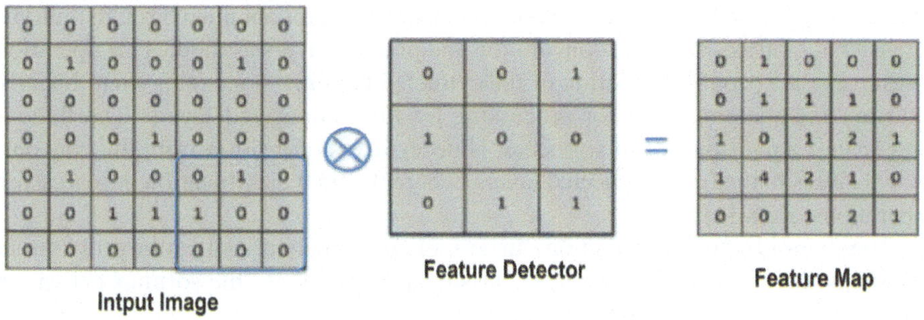

Fig. 2.4 Convolution to produce a feature map (adapted from Ajit et al. [1])

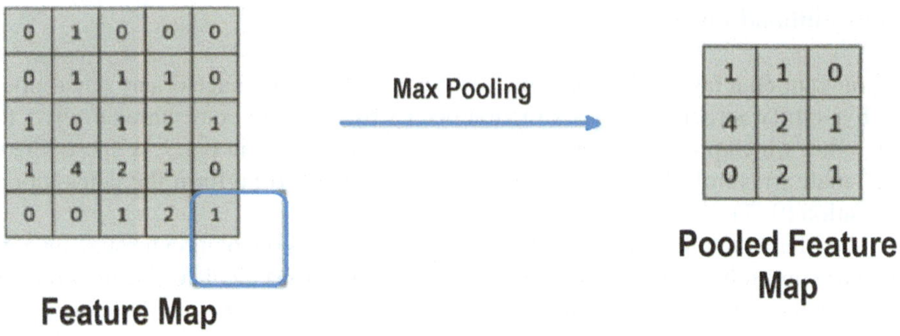

Fig. 2.5 Result of applying max pooling to a feature map (adapted from Ajit et al. [1])

Max pooling, as shown in Fig. 2.5, chooses the highest value of each submatrix of the feature maps resulting from subsequent layers and forms a new, separate matrix from these. This preserves dominant features and helps identify important patterns in the image. Max pooling is usually done using a 2×2 filter [1].

With average pooling, the procedure is similar. The difference is in how the final result is calculated, where the average of the sub-matrix is taken instead of highlighting the maximum value. This technique helps reduce the impact of noise and smooth information in regions [3].

Totally Connected

Artificial Neural Networks (ANNs) are computational processing systems heavily inspired by how the human brain's nervous system operates [11]. These networks comprise many interconnected computational nodes that collectively learn from the input to optimize the final output [11].

The last fully connected CNN layers perform the same tasks found in standard ANNs and attempt to produce scores for each class in a specific classification case. It is suggested that ReLu can be used in the intermediate layers to improve performance [11]. These layers receive a one-dimensional vector obtained through multidimensional matrices from the convolution and pooling layers as input. This operation preserves the spatial information of the extracted features, organizing them linearly and making them suitable for processing by fully connected layers. The flattening layer, responsible for this process, is necessary to transition from the convolutional layers to the fully connected layers, where the final classification or regression operations occur [1].

The sigmoid activation function can be applied to binary classification problems in the output layer. Regarding multiple classification problems, the softmax activation function is the most recommended [12].

References

1. Ajit, A., Acharya, K., Samanta, A.: A review of convolutional neural networks. In: 2020 International Conference on Emerging Trends in Information Technology and Engineering (IC-ETITE) (2020). https://doi.org/10.1109/ic-ETITE47903.2020.049
2. Arantes, M.: Método de reconhecimento da marcha humana por meio da fusão das características do movimento global (Doctoral dissertation). Universidade de São Paulo, São Paulo (2010)
3. Gu, J., Wang, Z., Kuen, J., Ma, L., Shahroudy, A., Shuai, B., Liu, T., Wang, X., Wang, G., Cai, J., Chen, T.: Recent advances in convolutional neural networks. Pattern Recogn. **77**, 354–377 (2018). https://doi.org/10.1016/j.patcog.2017.10.013
4. Gupta, S., Chattopadhyay, P.: Gait recognition in the presence of covariate conditions. Neurocomputing **454**, 76–87 (2021). https://doi.org/10.1016/j.neucom.2021.04.113
5. Halder, A., Chattopadhyay, P., Kumar, S.: Gait transformation network for gait re-identification with pose preservation. Signal Image Video Process. **17**(5), 1753–1761 (2023). https://doi.org/10.1007/s11760-022-02386-x
6. Harris, E., Khoo, I.-H., Demircan, E.: A survey of human gait-based artificial intelligence applications. Front. Robot. A **I**, 8 (2022). https://doi.org/10.3389/frobt.2021.749274
7. Hawas, A., El-Khobby, H., Elnaby, M., El-Samie, F.: Gait identification by convolutional neural networks and optical flow. Multimedia Tools Appl. **78**(18), 25873–25888 (2019). https://doi.org/10.1007/s11042-019-7638-9
8. Khaliluzzaman, M., Uddin, A., Deb, K., Hasan, M.J.: Person recognition based on deep gait: a survey. Sensors **23**(10) (2023). https://doi.org/10.3390/s23104875
9. Khan, A., Sohail, A., Zahoora, U., Qureshi, A.S.: A survey of the recent architectures of deep convolutional neural networks. Artif. Intell. Rev. **53**, 5455–5516 (2020). https://doi.org/10.1109/ICCTech55650.2022.00008
10. Khan, M.H., Farid, M.S., Grzegorzek, M.: A comprehensive study on codebook-based feature fusion for gait recognition. Inform. Fusion **92**, 216–230 (2022). https://doi.org/10.1016/j.inffus.2022.12.001
11. Li, Z., Liu, F., Yang, W., Peng, S., Zhou, J.: A survey of convolutional neural networks: analysis, applications, and prospects. IEEE Trans. Neural Netw. Learn. Syst. **33**(12), 6999–7019 (2022). https://doi.org/10.1109/TNNLS.2021.3084827
12. O'Shea, K., Nash, R.: An introduction to convolutional neural networks (2015)
13. Russel, N.S., Selvaraj, A.: Gender discrimination, age group classification and carried object recognition from gait energy image using fusion of parallel convolutional neural network. IET Image Process. **15**(1), 239–251 (2021). https://ietresearch.onlinelibrary.wiley.com/doi/pdf/10.1049/ipr2.12024, https://doi.org/10.1049/ipr2.12024
14. Sabilan, M., Aziz, A., Mohamad, F., Mansor, F., Hamid, N.: Leth-Gait: a fusion formula for gait-based gender classification using depth images. In: 2022 International Conference on Computer Technologies (ICCTech) (2022). https://doi.org/10.1109/ICCTech55650.2022.00008
15. Sharma, P., Khan, A., Jain, S., Srivastava, A.: Artificial intelligence driven human identification. J. Inform. Technol. Manag. **15**(3), 113–133 (2023). https://doi.org/10.22059/JITM.2023.93629
16. Soni, K., Dogra, D.P., Sekh, A.A., Kar, S., Choi, H., Kim, I.-J.: Person re-identification in indoor videos by information fusion using graph convolutional networks. Expert Syst. Appl. **210** (2022). https://doi.org/10.1016/j.eswa.2022.118363
17. Wang, L., Tan, T., Ning, H., Hu, W.: Silhouette analysis-based gait recognition for human identification. IEEE Trans. Pattern Anal. Mach. Intell. **25**, 1505–1518 (2004). https://doi.org/10.1109/TPAMI.2003.1251144

18. Xu, Z., Lu, W., Zhang, Q., Yeung, Y., Chen, X.: Gait recognition based on capsule network. J. Vis. Commun. Image Represent. **59**, 159–167 (2019). https://doi.org/10.1016/j.jvcir.2019.01.023
19. Yap, M.H., Pons, G., Martí, J., Ganau, S., Sentis, M., Zwiggelaar, R., Davison, A.K., Martí, R.: Automated breast ultrasound lesions detection using convolutional neural networks. IEEE J. Biomed. Health Inform. **22**(4), 1218–1226 (2018). https://doi.org/10.1109/JBHI.2017.2731873

Chapter 3
Research Objectives and Method

Abstract This chapter delineates the systematic review process to analyze computational vision-based techniques for gait-based identification and gender classification. The research question focuses on understanding these techniques' efficacy and implementation. A comprehensive search was conducted across the Scopus and Web of Science databases using a structured query, adhering to PRISMA guidelines. From an initial pool of over 3000 articles, 112 were initially considered, with 58 meeting the inclusion criteria after screening. Excluded articles were found based on thematic relationships, such as clinical gait analysis, wearable sensors, and emotion recognition. The exclusion criteria were refined to focus on gait identification with simple RGB images and exclude multi-camera setups, gender classification, and re-identification studies. Therefore, this chapter provides a detailed insight into the selection process, resulting in 56 relevant articles that form the basis of the systematic review's subsequent analysis, aiming to reflect the current state-of-the-art in the field.

3.1 Research Question

It was conducted a systematic review. This study started with a more generic search to find the most precise keywords to carry out the best possible search that led to the article [35]. The authors gathered more than 3000 articles from 2012 to mid-2021 that focus on studies of human gait and apply machine learning techniques. This article overviews six main applications of machine learning using gait data, informing of specific terms that characterize and distinguish them. Only one of these applications was of interest for this study: "Gait-based biometrics with applications in person identification, authentication, and re-identification as well as gender and age recognition" [35]. In the beginning, it was of interest the articles addressing computational systems capable of recognizing people by their gait and inferring their gender based on imaging sensors. It was questioned what the best search approach would be. Whether it would be better to carry out two independent pieces of search on person identification and gender recognition or to combine the two topics in a joint search. Reading this study, it was realized that they are related since the

D. R. M. Bastos and J. M. R. S. Tavares, *Advances in Gait-Based Identification*, Studies in Systems, Decision and Control 593,
https://doi.org/10.1007/978-3-031-89560-9_3

gender classification task is usually conducted alongside human identification and re-identification. They say the person's gender identification is done first to prune a subset of subjects before human identification is performed, serving as a soft feature in identifying a person [35].

To begin this systematic review, the research question the review aims to answer was defined. After a few drafts, it was concluded that the research aims to answer the question: "What are the computer vision-based techniques and methods for gait-based identification and gender classification in humans, and how can they be effectively implemented?".

3.2 Data Extration

To conduct a comprehensive search for published articles addressing the research question, it was performed searches in two well-known bibliographic databases: Scopus[1] and Web of Science.[2] Scopus is a widely recognized multidisciplinary bibliographic database maintained by Elsevier, a prominent publisher of scientific and technical journals. Web of Science, maintained by Clarivate Analytics, formerly a part of Thomson Reuters, is another highly regarded multidisciplinary bibliographic database. Using both Scopus and Web of Science databases ensures a comprehensive and well-rounded approach to this literature review, enabling the capture of a broad spectrum of relevant research findings and scholarly contributions. These two databases are pivotal in modern academic research, providing robust support for in-depth investigations and assessments of the scholarly landscape.

The search query was structured as follows: "gait AND (identif* OR recogni* OR biometric* OR "gender classification") AND ("machine learning" OR "deep learning" OR "computer vision" OR "neural architecture") AND (camera OR imag* OR vision OR frame OR frames)". The search query was designed to retrieve relevant articles that comprehensively address the research question and should only be applied to the title, abstract, and keywords of them.

The PRISMA[3] guidelines were followed in this systematic review. The PRISMA flow diagram is present in Fig. 3.1.

The number of studies has increased remarkably, demonstrating the research topic's growing interest and relevance. However, this substantial growth also challenges managing the volume of available literature. To solve this problem, it was decided to restrict the search to articles published in the last 5 years between 2019 and 2023. This data restriction is part of this systematic review's specific search limits to ensure the articles' relevance and timeliness. Furthermore, by restricting the search to more recent publications, the study is focused on the most recent research and developments in this field. This approach helped ensure the systematic review

[1] https://www.scopus.com.

[2] https://www.webofscience.com.

[3] http://www.prisma-statement.org.

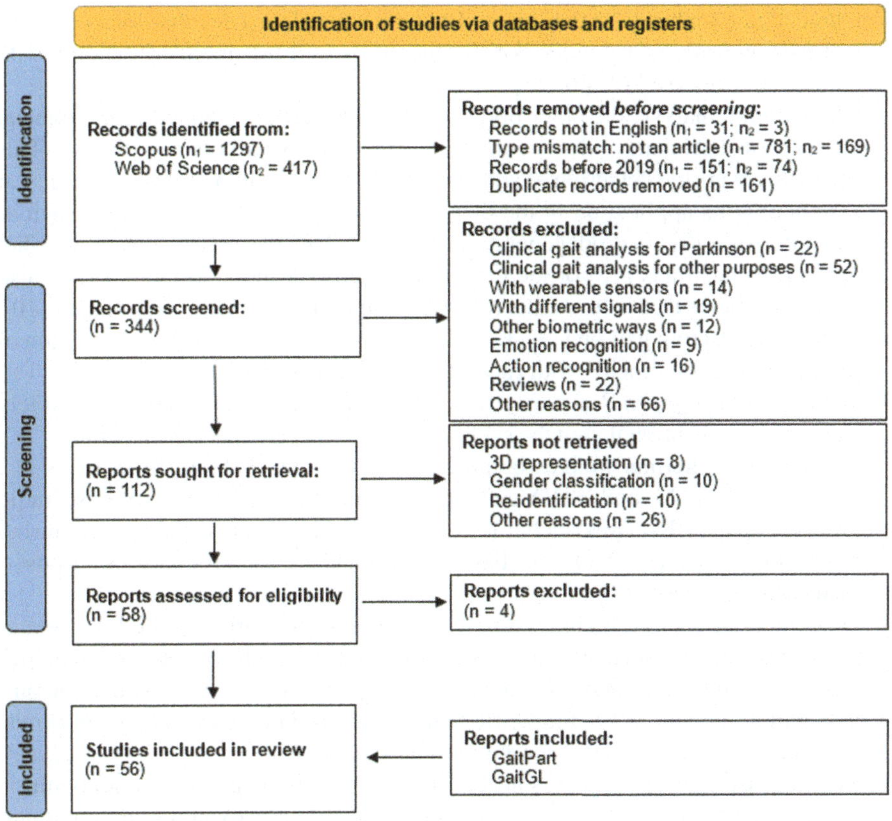

Fig. 3.1 PRISMA flow diagram

reflects the most up-to-date findings and knowledge. Considering the most current and relevant research is particularly important in this field, which is evolving rapidly as computational vision algorithms are involved. In addition, the articles with the most promising results are those published after 2020, as shown by a recent survey [47]. This review also only considered articles published in English.

After initial screening, the excluded articles were categorized by title, abstract, and keywords to provide transparency and traceability in the study selection process. Its categorization also allows the understanding of the framework of the main topic. Provides a comprehensive overview of the field and a more complete view of the knowledge available. Categories were selected based on criteria of relevance and similarity between articles, seeking to group studies that share thematic relationships or common characteristics. The categories chosen were the following:

1. Clinical gait analysis for Parkinson: Encompasses all articles that focus on clinical gait analysis using video for the diagnosis and treatment of patients with Parkinson's disease [12, 46, 77];
2. Clinical gait analysis for other purposes: Includes articles related to the clinical analysis of gait using video in contexts unrelated to Parkinson's disease. This category covers the detection of abnormal/pathological gait patterns. Of particular note here is its application in the diagnosis and treatment of conditions such as osteoporosis [49], neurodegenerative diseases [16], the assessment of alcohol levels in pedestrians [72] or the detection of fall risk in the elderly [19]. It also includes articles that help extract relevant pose points in characterizing gait [10]. Used for extracting gait phases [110], human pose estimation [1] or estimation of joint angles and moments with rehabilitation purposes [61];
3. With different signals: Refers to articles that acquire gait characteristics remotely using non-conventional signals, such as micro-Doppler radar spectrograms [18], silhouettes in thermal images obtained with infrared [37], Wi-Fi signals [11], LoRa signals [24], and a 3D gait motion analyzer employing impulse radio ultra-wideband (IR-UWB) technology [75]. Such signals are capable of acquiring movement in adverse conditions like low light and are used for various purposes, including gait recognition;
4. Other biometric ways: Includes articles addressing alternative methods for recognition beyond gait, such as identifying individuals through foot pressure data [6], footstep sounds [2], hand movements [65], or multimodal systems that encompass gait along with other biometric features like retina, faces, ears, palm prints, fingerprints, voice and DNA-based biometric data [27];
5. Emotion recognition: It covers articles related to automatic emotion recognition using vision-based methods in gait analysis. This category explores new computational methods for recognizing mental health and detecting the risk of depression and anxiety [114]. It also explores the use of deep learning algorithms to, for example, assess levels of happiness of pedestrians [50] and determine their levels of self-esteem [58];
6. Action recognition: It includes all articles on techniques for recognizing human actions and activities. Skeleton-based models to analyze human posture and detect video events such as abnormal human activity for surveillance systems [70]. Action recognition is also applied in more specific areas such as tennis [85] and soccer [53];
7. Reviews: References all the review articles. It includes reviews of interest on the main vision-based algorithms [71] and datasets used for gait recognition [79]. It also includes reviews of no interest to this study, such as state-of-the-art methods for the automatic detection of different brain abnormalities using magnetic resonance images [90] or multimodal methods (including gait analysis) for the diagnosis and treatment of patients with Parkinson [84];
8. Other reasons: Covers articles exploring diverse topics that do not fit into the previous categories, such as detecting musculoskeletal anomalies in X-ray images [5]. It also includes articles unrelated to humans, such as the assessment of the locomotion of broiler chickens [92], automatic lameness scoring of dairy cows [113],

and the analysis of pig gait patterns to identify swine fever [21]. Additionally, this category includes articles that do not directly answer the research question but are indirectly relevant, such as those addressing computational methods for testing the resilience of gait recognition models [59], studies on the performance and energy consumption of various CNN architectures on different hardware for gait recognition tasks [13], information about datasets [86], articles dealing with re-identification issues in gait for privacy concerns [33], and methods for generating selective adversarial examples that are correctly classified by a friendly gait recognition system and misclassified by an adversarial one [51].

During the systematic review process, several articles were identified which, although they did not directly fit the research question, contained information indirectly related to the topic. These articles were excluded from the main analysis.

With 112 reports sought for retrieval, there was a need to develop new, stricter exclusion criteria since analyzing such a large number of articles becomes impractical. It was well known that the main aim of this study was to study and compare the existing or originally proposed computational methods to achieve gait-based human recognition and classification from images, which was the basis for the research question. At this point, and after coming into contact with many relevant articles in the field, some more specific and alternative work directions became apparent.

Considering all the knowledge acquired so far, some new exclusion criteria were created to ensure the accuracy and relevance of the articles. These criteria were as follows:

1. 3D Representations: In conducting this systematic review, it was deliberately decided to direct it towards a more homogeneous set of investigations and approaches based on 2D representations. These can offer a broader and more applicable view in identifying people by gait. This is due, for example, to the fact that most conventional surveillance systems operate with cameras that acquire images in two dimensions. Therefore, studies focusing on extracting 3D features using Kinect sensors or similar technologies were excluded. This decision was based on several factors to optimize the review's relevance and efficiency. Firstly, the exclusion was based on the observation that, compared to datasets employing 2D representations, there is a more limited availability of datasets using 3D features from Kinect sensors. In the studies [42, 76], gait identification algorithms were created using 3D representations, and their performance was evaluated, in total, in six 3D skeleton-based gait datasets. Of these six, only two captured walking patterns from multiple points of view, but one contains only 20 individuals and the other 52 (with fewer angle variations), which is still too few. The scarcity of these datasets can impact the generalizability and validity of the results by restricting the diversity of contexts and conditions. Furthermore, 3D-based methods require multiple calibrated cameras, which is expensive and difficult to implement despite excellent results [97]. The greater computational effort associated with these technologies motivated the decision to exclude studies employing 3D sensors such as Kinect. Processing three-dimensional data can

require substantial computational resources, which can impact the practical viability and scalability of practical implementations;
2. Gender classification: Initially, the research planned to cover both gender recognition and gait identification, with the expectation of gaining valuable insights into possible interactions between these dimensions. However, after a preliminary review, it was found that there is a significant lack of work that comprehensively addresses both areas simultaneously. As gender recognition is not a central objective of the current study, it was decided to exclude articles that focus exclusively on gender classification [40], as well as age [116], weight and height estimates [88]. This decision aims to improve the efficiency of the review by focusing on the central aspects of gait identification. It is worth noting that although these specific articles were excluded from this review, it is recognized that they may contain valuable information, and consideration of their reading may be revised as necessary during the development of future developments, as inferring a person's gender could remain a secondary objective of related works;
3. Re-identification: The exclusion of works focusing on gait re-identification is motivated by several factors that differentiate this approach from gait identification in a system with a single camera. Among these factors, the scope of the task can be highlighted, where gait re-identification aims to recognize individuals at distinct locations and times, often using multiple cameras for continuous tracking [87]. In contrast, here, the focus is on recognizing identity based on gait in a specific context from a single camera. Furthermore, excluding articles involving multiple cameras aims to mitigate the additional computational complexity associated with gait re-identification in multi-camera scenarios. Challenges such as data fusion, camera matching, and temporal synchronization present difficulties that can compromise the effectiveness of systems. It is also important to note that gait re-identification often uses different datasets developed to deal with camera variations and environments [7, 106]. The decision to exclude these articles seeks to ensure consistency when considering algorithms and techniques applicable to specific datasets in the analysis focused on gait identification. This strategy does not prevent the consideration of relevant techniques for gait identification, even when they involve multiple cameras. On the contrary, it provides a more in-depth analysis adapted to this specific scope.

The full texts of the 112 articles sought for retrieval were then retrieved and subsequently assessed for eligibility, considering the three previously pre-defined exclusion criteria. Other documents did not fit this review's specific objectives and were excluded. As examples of this, it was found some articles that deal with gait identification for specific angles such as front view [112] or back view [117], evaluations in the confidentiality of the use of half gait with full gait [109], articles for which it was impossible to access the full text [48] or even articles with nothing related such as multimodal systems [74] or action recognition [52]. The excluded documents also include articles identifying conditions such as normal gait, carrying objects, or wearing a coat [62, 63, 82]. The term "gait recognition" is often used as a broad concept to describe individuals' identification or recognition based on specific gait characteristics. However, the specificity of what is considered "gait recognition" can vary between researchers and the contexts where the term is used. Some researchers may use the term to describe broader tasks, including identifying gait patterns under different conditions, such as normal walking, carrying objects, or wearing a coat. In these cases, the emphasis may be on the variation in gait pattern in response to different external factors. However, as the interest of this study lies exclusively in the unique identification of individuals based on their gait characteristics, it makes sense to exclude studies that focus solely on classification between different gait conditions, as these may not directly address the unique identification of each individual. Classification between different gait conditions is a different and less specific task compared to the unique identification of people based on their gait.

Having completed the screening process, the next phase involved a detailed examination of the full texts of the 58 remaining articles. During this stage, a comprehensive evaluation was conducted to extract detailed information from each article. The study design, methodology, datasets used, data analysis, methods, and outcomes were key factors to be considered. The goal was to gain a deeper understanding of the contributions of each report to the overall body of evidence. The results of this full-text review formed the basis for the final inclusion of reports in the systematic review.

After reading the full text, four articles were excluded based on distinct criteria that deviated from the primary focus on gait identification or met the exclusion criteria mentioned above. Each exclusion was driven by specific characteristics identified during the screening process. The four excluded articles were:

1. The first article proposed a novel method for human gait recognition, emphasizing precise segmentation and feature extraction at multiple levels. However, its primary focus was on recognizing normal gait, distinguishing gait with objects, and identifying individuals wearing a coat, rather than addressing the core objective of gait identification [81];
2. The second article presented an algorithm specifically engineered to withstand variations in human pose, body occlusion, and changes in camera angles. Employing sophisticated techniques such as image augmentation to enrich features, the algorithm was primarily oriented towards person re-identification. It extensively used the Market dataset, a large-scale public benchmark dataset tailored for person re-identification tasks. The emphasis on broader person re-identification challenges and datasets, while valuable in its own right, deviated from the specific context of gait identification that this systematic review aimed to address [78];
3. The third article explored Dynamic Vision Sensors (DVS) for human identification, capitalizing on their advantages in terms of low resource consumption and high temporal resolution. They use the polarity parameter provided by DVS, unique for capturing real-time intensity changes, distinguishing between brightness increase and decrease events. This information is absent in conventional RGB cameras. Tests are conducted on the CASIA-B dataset, which they converted to the event format using a sensor recording screen playbacks. Other datasets in the event format were acquired by them. The emphasis on DVS data and the strate-

gies employed for identification led to the exclusion of this methodology from this review, as it did not align with the primary goal of gait identification with simple RGB images [78];

4. The fourth article proposed an innovative real-time model-based gait analysis framework, incorporating a new Composite Filter Feature Selection (CFFS) and using 3D coordinates of joints extracted with MediaPipe. MediaPipe is an open-source machine-learning framework developed by Google. Despite its advanced features, the article leaned more towards re-identification systems, and the application scenario was tailored for multi-camera setups in areas like airport entrances and corridors [41].

Considering their consistent and noteworthy mentions across various articles within the systematic review, including two other articles in the final count is justifiable. These articles have emerged as significant references within the broader literature, underscoring their relevance to the overarching theme of gait identification. The articles include the approach of two gait recognition networks, named GaitPart [20] and GaitGL [56], which are discussed later. Incorporating these well-referenced articles is expected to ensure a comprehensive and representative analysis, drawing upon the insights and findings resonating throughout the scholarly discourse on gait identification.

In Table 3.1, it is presented the list of the 56 articles resulting from the entire process of inclusion and exclusion, i.e., the articles considered in this systematic review. Each reference indicates whether it adopts a model-based and/or model-free approach. Following the framework introduced by Khaliluzzaman et al. [47], it also included details about the employed neural network architecture. This information is a foundation for categorizing and comparing various articles, as discussed in Chap. 5. The gait feature representations and datasets used in each article are also specified, as these are critical points for providing an overview of the current state of the art. The articles are organized chronologically. Articles marked with the symbol "×" in the "Neural Network Architecture" column signify that the discussed method does not rely on DL. Similarly, articles with the symbol "×" in the "Dataset" column indicate that the chosen approach is tested and evaluated using independent datasets acquired by the authors.

The meticulous full-text review marked a critical step in ensuring the quality and relevance of the selected articles, contributing to the overall robustness of the systematic review.

Table 3.1 Summary of the final articles considered for the systematic review

References	Model based	Model free	Neural network architecture(s)	Gait feature representation	Dataset(s)
Xu et al. [104]		✓	CapsNet	GEI; CGI	CASIA-B; OU-ISIR Treadmill B
Babaee et al. [8]		✓	DAE	GEI	CASIA-B; OULP
Wang et al. [98]		✓	GAN	GEI	CASIA-B; OULP
Hawas et al. [36]		✓	CNN	Optical Flow (from GEI)	CASIA-B
Thapar et al. [89]		✓	3D-CNN + LSTM	SSM	CASIA-B; OULP
Zhang et al. [111]		✓	CNN + LSTM	Silhouettes	CASIA-B; OULP; OUMVLP
Linda et al. [57]		✓	CNN	CCGI	CASIA-B
Wu et al. [101]		✓	×	2D-EGEI	CASIA-B; OULP
Xue et al. [105]		✓	GAN + CNN	GEI	CASIA-B; OULP
Wang and Yan [95]		✓	CNN + LSTM	ff-GEI	CASIA-B; OULP
Wang and Zhang [96]		✓	CNN	MF-GEI	CASIA-B; OULP
Li et al. [54]		✓	DAE + CNN	GEI	OU-LP-Bag; OU-LP-Bag β; OU-ISIR Treadmill B; TUM-GAID
Castro et al. [14]		✓	3D-CNN; 2D-CNN	(Gray-level Pixels + Optical Flow + Depth Maps) from raw input data	CASIA-B; TUM-GAID
Guo et al. [29]		✓	×	GEI	CASIA-A; CASIA-B
George et al. [25]		✓	CNN	CCGI	CASIA-B; OULP
Wang et al. [97]		✓	CNN	Silhouettes	CASIA-B; OULP
Fan et al. [20]		✓	CNN	Silhouettes	CASIA-B; OUMVLP
Bekhouch et al. [9]	✓		×	skeleton	CASIA-B
Moghaddam and Etemad [69]		✓	RNN	GCEM	CASIA-B; OUMVLP
Gupta and Chattopadhyay [30]		✓	DAE	GAEI	CASIA-B; TUM-GAID

(continued)

Table 3.1 (continued)

References	Model based	Model free	Neural network architecture(s)	Gait feature representation	Dataset(s)
Lima et al. [55]	✓		×	skeleton (OpenPose format)	CASIA-A; CASIA-B
Wu et al. [102]		✓	DAE	Silhouettes	CASIA-B; OULP
Chao et al. [15]		✓	CNN	Silhouettes	CASIA-B; OUMVLP
Jameel and Dhannoon [44]		✓	CNN	Silhouettes	×
Gul et al. [28]		✓	3D-CNN	colored GEI	CASIA-B; OULP
Wen and Wang [100]		✓	×	ff-GEI	CASIA-B; OULP
Gupta and Chattopadhyay [31]		✓	GAN	DGEI	CASIA-B; TUM-GAID; OU-ISIR Treadmill B
Wen and Wang [99]		✓	CNN + LSTM	ff-GEI	CASIA-B; OULP
Xu et al. [103]	✓		CNN	skeleton	CASIA-A; CASIA-B; CASIA-C
Jia et al. [45]		✓	CNN + ViT	Silhouettes contour	CASIA-A; CASIA-B
Iwashita et al. [43]		✓	×	GEI	CASIA-C; OU-ISIR Treadmill A; GaitST
Gao et al. [22]	✓		GCN	skeleton (OpenPose format)	CASIA–B; OUMVLP–Pose
Alsaggaf et al. [4]		✓	GAN + CNN	GEI	CASIA-B
Merlin Linda et al. [64]		✓	CNN + CapsNet	CCGI	CASIA-B
Marin-Jimenez et al. [60]		✓	CNN	(Gray-level Pixels + Optical Flow + Depth Maps) from raw input data; silhouettes	CASIA-B; TUM-GAID
Han et al. [34]		✓	CNN	Silhouettes	CASIA-B; TUM-GAID
Hou et al. [38]		✓	DAE	Silhouettes	CASIA-B; OUMVLP
Wang and Hu [94]		✓	CNN	ff-GEI	CASIA-B; OUMVLP

(continued)

Table 3.1 (continued)

References	Model based	Model free	Neural network architecture(s)	Gait feature representation	Dataset(s)
Yao et al. [107]		✓	DAE	Silhouettes	CASIA-B; OU-ISIR Treadmill B
Zheng et al. [115]	✓		GCN	skeleton (HRNet format)	CASIA-B
Gao et al. [23]		✓	CNN	Silhouettes	CASIA-B; OU-ISIR Treadmill B
Ghosh [26]		✓	LSTM	ff-GEI	CASIA-B; OU-LP-BAG; OU-ISIR Treadmill B; OULP-Age
Mogan et al. [67]		✓	CNN	GEI	CASIA-B; OULP; OU-ISIR Treadmill D
Lin et al. [56]		✓	3D-CNN	Silhouettes	CASIA-B; OUMVLP; GREW; Gait3D
Sayeed et al. [80]		✓	CNN	GEI	OUMVLP; OULP
Pinčić et al. [73]		✓	ViT	GEI	CASIA-B; OUMVLP
Mogan et al. [68]		✓	ViT	GEI	CASIA-B; OULP; OU-ISIR Treadmill D
Upadhyay and Gonsalves [91]	✓		RNN	skeleton (OpenPose format)	CASIA-B
Aljazaerly et al. [3]		✓	CNN	GEI	CASIA-B; OUMVLP
Hua et al. [39]	✓	✓	CNN	Silhouettes + skeleton (OpenPose format)	CASIA-B
Wang et al. [93]		✓	CNN	PEI	CASIA-B; OUMVLP
Cosma et al. [17]	✓		ViT	skeleton (COCO format)	ASIA-B; GREW; DenseGait
Mogan et al. [66]		✓	CNN + ViT	GEI	CASIA-B; OULP; OU-ISIR Treadmill D
Yeo et al. [108]		✓	CNN	GEI	CASIA-B
Sharma et al. [83]		✓	×	Silhouettes	×
Gupta and Chattopadhyay [32]		✓	GAN	BEI	CASIA-B; TUM-GAID

References

1. Åberg, A.C., Olsson, F., Åhman, H.B., Tarassova, O., Arndt, A., Giedraitis, V., Berglund, L., Halvorsen, K.: Extraction of gait parameters from marker-free video recordings of Timed Up-and-Go tests: validity, inter- and intra-rater reliability. Gait Posture **90**, 489–495 (2021). https://doi.org/10.1016/j.gaitpost.2021.08.004
2. Algermissen, S., Hornlein, M.: Person identification by footstep sound using convolutional neural networks. Appl. Mech. **2**(2), 257–273 (2021). https://doi.org/10.3390/applmech2020016
3. Aljazaerly, M.A.A., Makihara, Y., Muramatsu, D., Yagi, Y.: Batch hard contrastive loss and its application to cross-view gait recognition. IEEE Access **11**, 31177–31187 (2022). https://doi.org/10.1109/ACCESS.2023.3262271
4. Alsaggaf, W.A., Mehmood, I., Khairullah, E.F., Alhuraiji, S., Sabir, M.F.S., Alghamdi, A.S., Abd El-Latif, A.A.: A smart surveillance system for uncooperative gait recognition using cycle consistent generative adversarial networks (CCGANs). Comput. Intell. Neurosci. (2021). https://doi.org/10.1155/2021/3110416
5. Amiri, S., Akbarabadi, M., Rimaz, S., Abdolali, F., Ahadi, R., Afshani, M., Askarabad, Z.A., Kowsarirad, T., Sakinehpour, S., Ayvazzadeh, N., Cheraghi, S.: Designing an intelligent lesion detection system using deep architecture neural networks in the lower limb X-ray images. Front. Biomed. Technol. **10**(2), 169–179 (2023). https://doi.org/10.18502/fbt.v10i2.12221
6. Ardhianto, P., Subiakto, R.B.R., Lin, C.-Y., Jan, Y.-K., Liau, B.-Y., Tsai, J.-Y., Akbari, V.B.H., Lung, C.-W.: A deep learning method for foot progression angle detection in plantar pressure images. Sensors **22**(7) (2022). https://doi.org/10.3390/s22072786
7. Avola, D., Cinque, L., Fagioli, A., Foresti, G.L., Pannone, D., Piciarelli, C.: Bodyprint–a meta-feature based LSTM hashing model for person re-identification. Sensors (Switzerland) **20**(18), 1–19 (2020). https://doi.org/10.3390/s20185365
8. Babaee, M., Li, L., Rigoll, G.: Person identification from partial gait cycle using fully convolutional neural networks. Neurocomputing **338**, 116–125 (2019). https://doi.org/10.1016/j.neucom.2019.01.091
9. Bekhouch, A., Bouchrika, I., Doghmane, N.: Gait biometrics: investigating the use of the lower inner regions for people identification from landmark frames. IET Biometrics **9**(6), 269–277 (2020). https://doi.org/10.1049/iet-bmt.2020.0001
10. Benenaula, S., Trelles, M.D., Garza-Castañón, L.E., Minchala, L.I.: Classification of gait anomalies by using space-time parameters obtained with pose estimation. Int. J. Innov. Comput. Inform. Control **18**(6), 1913–1927 (2022). https://doi.org/10.24507/ijicic.18.06.1913
11. Bu, Q., Ming, X., Hu, J., Zhang, T., Feng, J., Zhang, J.: TransferSense: towards environment independent and one-shot wifi sensing. Pers. Ubiquit. Comput. **26**(3), 555–573 (2022). https://doi.org/10.1007/s00779-020-01480-6
12. Buongiorno, D., Bortone, I., Cascarano, G.D., Trotta, G.F., Brunetti, A., Bevilacqua, V.: A low-cost vision system based on the analysis of motor features for recognition and severity rating of Parkinson's disease. BMC Med. Inform. Decis. Mak. **19**(9) (2019). https://doi.org/10.1186/s12911-019-0987-5
13. Castro, F.M., Guil, N., Marin-Jimenez, M.J., Perez-Serrano, J., Ujaldon, M.: Energy-based tuning of convolutional neural networks on multi-GPUs. Concurr. Comput.-Pract. Experience **31**(21, SI) (2019). https://doi.org/10.1002/cpe.4786
14. Castro, F., Marín-Jiménez, M., Guil, N., Blanca, N.: Multimodal feature fusion for CNN-based gait recognition: an empirical comparison. Neural Comput. Appl. **32**(17), 14173–14193 (2020). https://doi.org/10.1007/s00521-020-04811-z
15. Chao, H., Wang, K., He, Y., Zhang, J., Feng, J.: GaitSet: cross-view gait recognition through utilizing gait as a deep set. IEEE Trans. Pattern Anal. Mach. Intell. **44**(7), 3467–3478 (2021). https://doi.org/10.1109/TPAMI.2021.3057879
16. Cheriet, M., Dentamaro, V., Hamdan, M., Impedovo, D., Pirlo, G.: Multi-speed transformer network for neurodegenerative disease assessment and activity recognition. Comput. Methods Programs Biomed. **230** (2023). https://doi.org/10.1016/j.cmpb.2023.107344

17. Cosma, A., Catruna, A., Radoi, E.: Exploring self-supervised vision transformers for gait recognition in the wild. Sensors **23**(5) (2023). https://doi.org/10.3390/s23052680
18. Doherty, H.G., Burgueño, R.A., Trommel, R.P., Papanastasiou, V., Harmanny, R.I.A.: Attention-based deep learning networks for identification of human gait using radar micro-Doppler spectrograms. Int. J. Microwave Wirel. Technol. **13**(7), 734–739 (2021). https://doi.org/10.1017/S1759078721000830
19. Dubois, A., Bihl, T., Bresciani, J.-P.: Identifying fall risk predictors by monitoring daily activities at home using a depth sensor coupled to machine learning algorithms. Sensors **21**(6), 1–10 (2021). https://doi.org/10.3390/s21061957
20. Fan, C., Peng, Y., Cao, C., Liu, X., Hou, S., Chi, J., Huang, Y., Li, Q., He, Z.: Gaitpart: temporal part-based model for gait recognition. In: 2020 IEEE/CVF Conference on Computer Vision and Pattern Recognition (CVPR), pp. 14213–14221 (2020). https://doi.org/10.1109/CVPR42600.2020.01423
21. Fernández-Carrión, E., Barasona, J.Á., Sánchez, Á., Jurado, C., Cadenas-Fernández, E., Sánchez-Vizcaíno, J.M.: Computer vision applied to detect lethargy through animal motion monitoring: a trial on African swine fever inwild boar. Animals **10**(12), 1–12 (2020). https://doi.org/10.3390/ani10122241
22. Gao, S., Yun, J., Zhao, Y., Liu, L.: Gait-D: skeleton-based gait feature decomposition for gait recognition. IET Comput. Vis. **16**(2), 111–125 (2021). https://doi.org/10.1049/cvi2.12070
23. Gao, Z., Wu, J., Wu, T., Huang, R., Zhang, A., Zhao, J.: Robust clothing-independent gait recognition using hybrid part-based gait features. PeerJ Comput. Sci. **8** (2022). https://doi.org/10.7717/peerj-cs.996
24. Ge, Y., Li, W., Farooq, M., Qayyum, A., Wang, J., Chen, Z., Cooper, J., Imran, M.A., Abbasi, Q.H.: LoGait: LoRa sensing system of human gait recognition using dynamic time warping. IEEE Sens. J. **23**(18), 21687–21697 (2023). https://doi.org/10.1109/JSEN.2023.3297438
25. George, M., Govindarajan, T., Rajasekaran, K., Bandi, S.: A robust similarity based deep siamese convolutional neural network for gait recognition across views. Comput. Intell. **36**(3), 1290–1319 (2020). https://doi.org/10.1111/coin.12361
26. Ghosh, R.: A Faster R-CNN and recurrent neural network based approach of gait recognition with and without carried objects. Expert Syst. Appl. **205** (2022). https://doi.org/10.1016/j.eswa.2022.117730
27. Gona, A., Subramoniam, M., Swarnalatha, R.: Transfer learning convolutional neural network with modified Lion optimization for multimodal biometric system. Comput. Electr. Eng. **108** (2023). https://doi.org/10.1016/j.compeleceng.2023.108664
28. Gul, S., Malik, M.I., Khan, G.M., Shafait, F.: Multi-view gait recognition system using spatio-temporal features and deep learning. Expert Syst. Appl. **179** (2021). https://doi.org/10.1016/j.eswa.2021.115057
29. Guo, H., Li, B., Zhang, Y., Zhang, Y., Li, W., Qiao, F., Rong, X., Zhou, S.: Gait recognition based on the feature extraction of gabor filter and linear discriminant analysis and improved local coupled extreme learning machine. Math. Prob. Eng. **2020** (2020). https://doi.org/10.1155/2020/5393058
30. Gupta, S., Chattopadhyay, P.: Exploiting pose dynamics for human recognition from their gait signatures. Multimedia Tools Appl. **80**(28–29), 35903–35921 (2020). https://doi.org/10.1007/s11042-020-10071-9
31. Gupta, S., Chattopadhyay, P.: Gait recognition in the presence of co-variate conditions. Neurocomputing **454**, 76–87 (2021). https://doi.org/10.1016/j.neucom.2021.04.113
32. Gupta, S., Chattopadhyay, P.: Pose-based boundary energy image for gait recognition from silhouette contours. Sadhana—Acad. Proc. Eng. Sci. **48**(4) (2023). https://doi.org/10.1007/s12046-023-02234-9
33. Halder, A., Chattopadhyay, P., Kumar, S.: Gait transformation network for gait re-identification with pose preservation. Signal Image Video Process. **17**(5), 1753–1761 (2023). https://doi.org/10.1007/s11760-022-02386-x
34. Han, F., Li, X., Zhao, J., Shen, F.: A unified perspective of classification-based loss and distance-based loss for cross-view gait recognition. Pattern Recogn. **125** (2022). https://doi.org/10.1016/j.patcog.2021.108519

35. Harris, E., Khoo, I.-H., Demircan, E.: A survey of human gait-based artificial intelligence applications. Front. Robot. AI **8** (2022). https://doi.org/10.3389/frobt.2021.749274
36. Hawas, A., El-Khobby, H., Elnaby, M., El-Samie, F.: Gait identification by convolutional neural networks and optical flow. Multimedia Tools Appl. **78**(18), 25873–25888 (2019). https://doi.org/10.1007/s11042-019-7638-9
37. Hossain, M.S.B., Dranetz, J., Choi, H., Guo, Z.: DeepBBWAE-Net: a CNN-RNN based deep SuperLearner for estimating lower extremity sagittal plane joint kinematics using shoe-mounted IMU sensors in daily living. IEEE J. Biomed. Health Inform. **26**(8), 3906–3917 (2022). https://doi.org/10.1109/JBHI.2022.3165383
38. Hou, S., Liu, X., Cao, C., Huang, Y.: Gait quality aware network: toward the interpretability of silhouette-based gait recognition. IEEE Trans. Neural Netw. Learn. Syst. **34**(11), 8978–8988 (2022). https://doi.org/10.1109/TNNLS.2022.3154723
39. Hua, C., Pan, Y., Li, J., Wang, Z.: Gait recognition by combining the long-short-term attention network and personal physiological features. Sensors **22**(22) (2022). https://doi.org/10.3390/s22228779
40. Huang, B., Luo, Y., Xie, J., Pan, J., Zhou, C.: Attention-aware spatio-temporal learning for multi-view gait-based age estimation and gender classification. IET Comput. Vis. (2022). https://doi.org/10.1049/cvi2.12165
41. Huang, X., Dong, R., Wu, B., Sato, K., Ikuno, S., Wang, Z., Nishimura, S.: A real-time recognition gait framework for personal authentication via image-based neural network: accelerated by feature reduction in time and frequency domains. J. Real-Time Image Process. **20**(5) (2023). https://doi.org/10.1007/s11554-023-01349-w
42. Huynh-The, T., Hua, C.-H., Tu, N.A., Kim, D.-S.: Learning 3D spatiotemporal gait feature by convolutional network for person identification. Neurocomputing **397**, 192–202 (2020). https://doi.org/10.1016/j.neucom.2020.02.048
43. Iwashita, Y., Sakano, H., Kurazume, R., Stoica, A.: Speed invariant gait recognition—the enhanced mutual subspace method. PLoS ONE **16**(8 August) (2021). https://doi.org/10.1371/journal.pone.0255927
44. Jameel, H., Dhannoon, B.: Gait recognition based on deep learning. Iraqi J. Sci. **63**(1), 397–408 (2021). https://doi.org/10.24996/ijs.2022.63.1.36
45. Jia, P., Zhao, Q., Li, B., Zhang, J.: CJAM: Convolutional neural network joint attention mechanism in gait recognition. IEICE Trans. Inform. Syst. **E104D**(8), 1239–1249 (2021). https://doi.org/10.1587/transinf.2020BDP0010
46. Jin, C., Qi, S., Yang, L., Teng, Y., Li, C., Yao, Y., Ruan, X., Wei, X.: Abnormal functional connectivity density involvement in freezing of gait and its application for subtyping Parkinson's disease. Brain Imaging Behav. **17**(4), 375–385 (2023). https://doi.org/10.1007/s11682-023-00765-7
47. Khaliluzzaman, M., Uddin, A., Deb, K., Hasan, M.J.: Person recognition based on deep gait: a survey. Sensors **23**(10) (2023). https://doi.org/10.3390/s23104875
48. Khan, A., Fathima, S.: Multi gait recognition using clustering based faster regions-convolutional neural network. J. Intell. Fuzzy Syst. **44**(5), 8597–8606 (2023). https://doi.org/10.3233/JIFS-224114
49. Khan, M., Kadry, S., Parwekar, P., Damaševičius, R., Mehmood, A., Khan, J., Naqvi, S.: Human gait analysis for osteoarthritis prediction: a framework of deep learning and kernel extreme learning machine. Comp. Intell. Syst. **9**(3), 2665–2683 (2023). https://doi.org/10.1007/s40747-020-00244-2
50. Kim, J., Nirjhar, E.H., Kim, J., Chaspari, T., Ham, Y., Winslow, J.F., Lee, C., Ahn, C.R.: Capturing environmental distress of pedestrians using multimodal data: the interplay of biosignals and image-based data. J. Comput. Civil Eng. **36**(2) (2022). https://doi.org/10.1061/(ASCE)CP.1943-5487.0001009
51. Kwon, H.: Toward selective adversarial attack for gait recognition systems based on deep neural network. IEICE Trans. Inform. Syst. **E106D**(2), 262–266 (2023). https://doi.org/10.1587/transinf.2021EDL8080

52. Leyva, R., Sanchez, V., Li, C.-T.: Compact and low-complexity binary feature descriptor and fisher vectors for video analytics. IEEE Trans. Image Process. **28**(12), 6169–6184 (2019). https://doi.org/10.1109/TIP.2019.2922826
53. Li, R., Bhanu, B.: Energy-motion features aggregation network for players' fine-grained action analysis in soccer videos. IEEE Trans. Circ. Syst. Video Technol. 1–1 (2023). https://doi.org/10.1109/TCSVT.2023.3288565
54. Li, X., Makihara, Y., Xu, C., Yagi, Y., Ren, M.: Joint intensity transformer network for gait recognition robust against clothing and carrying status. IEEE Trans. Inform. Forensics Secur. **14**(12), 3102–3115 (2019). https://doi.org/10.1109/TIFS.2019.2912577
55. Lima, V., Melo, V., Schwartz, W.: Simple and efficient pose-based gait recognition method for challenging environments. Pattern Anal. Appl. **24**(2), 497–507 (2020). https://doi.org/10.1007/s10044-020-00935-z
56. Lin, B., Zhang, S., Wang, M., Li, L., Yu, X.:. GaitGL: learning discriminative global-local feature representations for gait recognition (2022)
57. Linda, G.M., Themozhi, G., Bandi, S.R.: Color-mapped contour gait image for cross-view gait recognition using deep convolutional neural network. Int. J. Wavelets Multiresolut. Inform. Process. **18**(1) (2019). https://doi.org/10.1142/S0219691319410121
58. Liu, X., Wen, Y., Zhu, T.: Ecological recognition of self-esteem leveraged by video-based gait. Front. Psychiat. **13** (2022). https://doi.org/10.3389/fpsyt.2022.1027445
59. Maqsood, M., Yasmin, S., Gillani, S., Aadil, F., Mehmood, I., Rho, S., Yeo, S.-S.: An autonomous decision-making framework for gait recognition systems against adversarial attack using reinforcement learning. ISA Trans. **132**, 80–93 (2023). https://doi.org/10.1016/j.isatra.2022.11.016
60. Marin-Jimenez, M.J., Castro, F.M., Delgado-Escano, R., Kalogeiton, V., Guil, N.: UGaitNet: multimodal gait recognition with missing input modalities. IEEE Trans. Inform. Forensics Secur. **16**, 5452–5462 (2021). https://doi.org/10.1109/TIFS.2021.3132579
61. Martín, G.D.-S., Reyes-González, L., Sainz-Ruiz, S., Rodríguez-Cobo, L., López-Higuera, J.M.: Automatic ankle angle detection by integrated RGB and depth camera system. Sensors **21**(5), 1–21 (2021). https://doi.org/10.3390/s21051909
62. Mehmood, A., Khan, M.A., Sharif, M., Khan, S.A., Shaheen, M., Saba, T., Riaz, N., Ashraf, I.: Prosperous human gait recognition: an end-to-end system based on pre-trained CNN features selection. Multimedia Tools Appl. (2020). https://doi.org/10.1007/s11042-020-08928-0
63. Mehmood, A., Khan, M.A., Tariq, U., Jeong, C.-W., Nam, Y., Mostafa, R.R., ElZeiny, A.: Human gait recognition: a deep learning and best feature selection framework. Comput. Mater. Continua **70**(1), 343–360 (2021). https://doi.org/10.32604/cmc.2022.019250
64. Merlin Linda, G., Sree Rathna Lakshmi, N.V.S., Murugan, N.S., Mahapatra, R.P., Muthukumaran, V., Sivaram, M.: Intelligent recognition system for viewpoint variations on gait and speech using CNN-CapsNet. Int. J. Intell. Comput. Cybern. **15**(3), 363–382 (2021). https://doi.org/10.1108/IJICC-08-2021-0178
65. Miyazaki, R., Sasaki, K., Tsumura, N., Hirai, K.: RGB-D video based hand authentication using deep neural network†. J. Imaging Sci. Technol. **67**(3) (2023). https://doi.org/10.2352/J.ImagingSci.Technol.2023.67.3.030502
66. Mogan, J., Lee, C., Lim, K., Ali, M., Alqahtani, A.: Gait-CNN-ViT: multi-model gait recognition with convolutional neural networks and vision transformer. Sensors **23**(8) (2023). https://doi.org/10.3390/s23083809
67. Mogan, J., Lee, C., Lim, K., Muthu, K.: VGG16-MLP: gait recognition with fine-tuned VGG-16 and multilayer perceptron. Appl. Sci. (Switzerl.) **12**(15) (2022). https://doi.org/10.3390/app12157639
68. Mogan, J.N., Lee, C.P., Lim, K.M., Muthu, K.S.: Gait-ViT: gait recognition with vision transformer. Sensors **22**(19) (2022). https://doi.org/10.3390/s22197362
69. Moghaddam, A., Etemad, A.: View-invariant gait recognition with attentive recurrent learning of partial representations. IEEE Trans. Biometrics Behav. Identity Sci. **3**(1), 124–137 (2020). https://doi.org/10.1109/TBIOM.2020.3031470

70. Monica, K.M., Parvathi, R.: Efficient gait analysis using deep learning techniques. Comput. Mater. Continua **74**(3), 6229–6249 (2023). https://doi.org/10.32604/cmc.2023.032273
71. Parashar, A., Parashar, A., Ding, W., Shekhawat, R.S., Rida, I.: Deep learning pipelines for recognition of gait biometrics with covariates: a comprehensive review. Artif. Intell. Rev. **56**(8), 8889–8953 (2023). https://doi.org/10.1007/s10462-022-10365-4
72. Park, S., Bae, B., Kang, K., Kim, H., Nam, M. S., Um, J., Heo, Y.J.: A deep-learning approach for identifying a drunk person using gait recognition. Appl. Sci. (Switzerl.) **13**(3) (2023). https://doi.org/10.3390/app13031390
73. Pinčić, D., Sušanj, D., Lenac, K.: Gait recognition with self-supervised learning of gait features based on vision transformers. Sensors **22**(19) (2022). https://doi.org/10.3390/s22197140
74. Rajasab, N., Rafi, M.: A deep learning approach for biometric security in video surveillance system using gait. Int. J. Saf. Secur. Eng. **12**(4), 491–499 (2022). https://doi.org/10.18280/ijsse.120410
75. Rana, S.P., Dey, M., Ghavami, M., Dudley, S.: Markerless gait classification employing 3D IR-UWB physiological motion sensing. IEEE Sens. J. **22**(7), 6931–6941 (2022). https://doi.org/10.1109/JSEN.2022.3154092
76. Rashmi, M., Guddeti, R.M.R.: Exploiting skeleton-based gait events with attention-guided residual deep learning model for human identification. Appl. Intell. (2023). https://doi.org/10.1007/s10489-023-05019-z
77. Sabo, A., Mehdizadeh, S., Ng, K.-D., Iaboni, A., Taati, B.: Assessment of Parkinsonian gait in older adults with dementia via human pose tracking in video data. J. NeuroEng. Rehabil. **17**(1) (2020). https://doi.org/10.1186/s12984-020-00728-9
78. Saleh, A.M., Hamoud, T.: Analysis and best parameters selection for person recognition based on gait model using CNN algorithm and image augmentation. J. Big Data **8**(1) (2021). https://doi.org/10.1186/s40537-020-00387-6
79. Santos, C., Oliveira, D., Passos, L., Pires, R., Santos, D., Valem, L., Moreira, T., Santana, M., Roder, M., Papa, J., Colombo, D.: Gait recognition based on deep learning: a survey. ACM Comput. Surv. **55**(2) (2022). https://doi.org/10.1145/3490235
80. Sayeed, M.S., Min, P.P., Bari, M.A.: Deep learning based gait recognition using convolutional neural network in the COVID-19 pandemic. Emerg. Sci. J. **6**(5), 1086–1099 (2022). https://doi.org/10.28991/ESJ-2022-06-05-012
81. Sharif, M., Attique, M., Tahir, M.Z., Yasmim, M., Saba, T., Tanik, U.J.: A machine learning method with threshold based parallel feature fusion and feature selection for automated gait recognition. J. Organ. End User Comput. **32**(2), 67–92 (2020). https://doi.org/10.4018/JOEUC.2020040104
82. Sharif, M.I., Khan, M.A., Alqahtani, A., Nazir, M., Alsubai, S., Binbusayyis, A., Damaševičius, R.: Deep learning and kurtosis-controlled, entropy-based framework for human gait recognition using video sequences. Electron. (Switzerl.) **11**(3) (2022). https://doi.org/10.3390/electronics11030334
83. Sharma, P., Khan, A., Jain, S., Srivastava, A.: Artificial intelligence driven human identification. J. Inform. Technol. Manag. **15**(3), 113–133 (2023). https://doi.org/10.22059/JITM.2023.93629
84. Skaramagkas, V., Pentari, A., Kefalopoulou, Z., Tsiknakis, M.: Multi-modal deep learning diagnosis of Parkinson's disease—a systematic review. IEEE Trans. Neural Syst. Rehabil. Eng. **31**, 2399–2423 (2023). https://doi.org/10.1109/TNSRE.2023.3277749
85. Skublewska, M., Powroznik, P.: Temporal pattern attention for multivariate time series of tennis strokes classification. Sensors **23**(5) (2023). https://doi.org/10.3390/s23052422
86. Song, C., Huang, Y., Wang, W., Wang, L.: CASIA-E: a large comprehensive dataset for gait recognition. IEEE Trans. Pattern Anal. Mach. Intell. **45**(3), 2801–2815 (2023). https://doi.org/10.1109/TPAMI.2022.3183288
87. Soni, K., Dogra, D.P., Sekh, A.A., Kar, S., Choi, H., Kim, I.-J.: Person re-identification in indoor videos by information fusion using graph convolutional networks. Expert Syst. Appl. **210** (2022). https://doi.org/10.1016/j.eswa.2022.118363

88. Taha, K., Yoo, P.D., Al-Hammadi, Y., Muhaidat, S., Yeun, C.Y.: Learning a deep-feature clustering model for gait-based individual identification. Comput. Secur.**136** (2023). https://doi.org/10.1016/j.cose.2023.103559
89. Thapar, D., Jaswal, G., Nigam, A., Arora, C.: Gait metric learning Siamese network exploiting dual of spatio-temporal 3D-CNN intra and LSTM based inter gait-cycle-segment features. Pattern Recogn. Lett. **125**, 646–653 (2019). https://doi.org/10.1016/j.patrec.2019.07.008
90. Tiwari, A., Srivastava, S., Pant, M.: Brain tumor segmentation and classification from magnetic resonance images: review of selected methods from 2014 to 2019. Pattern Recogn. Lett. **131**, 244–260 (2020). https://doi.org/10.1016/j.patrec.2019.11.020
91. Upadhyay, J., Gonsalves, T.: An enhanced gait recognition system based on the features fusion methodology with recurrent neural network (RNN). Indian J. Comput. Sci. Eng. **13**(5), 1483–1496 (2022). https://doi.org/10.21817/indjcse/2022/v13i5/221305042
92. van der Eijk, J.A.J., Guzhva, O., Schulte-Landwehr, J., Giersberg, M.F., Jacobs, L., de Jong, I.C.: Individuality of a group: detailed walking ability analysis of broiler flocks using optical flow approach. Smart Agric. Technol. **5** (2023). https://doi.org/10.1016/j.atech.2023.100298
93. Wang, R., Shi, Y., Ling, H., Li, Z., Li, P., Liu, B., Zheng, H., Wang, Q.: Gait recognition via gait period set. IEEE Trans. Biometrics Behav. Identity Sci. **5**(2), 183–195 (2023). https://doi.org/10.1109/TBIOM.2023.3244206
94. Wang, X., Hu, S.: Visual gait recognition based on convolutional block attention network. Multimedia Tools Appl. **81**(20), 29459–29476 (2022). https://doi.org/10.1007/s11042-022-12831-1
95. Wang, X., Yan, W.Q.: Human gait recognition based on frame-by-frame gait energy images and convolutional long short-term memory. Int. J. Neural Syst. **30**(1) (2019). https://doi.org/10.1142/S0129065719500278
96. Wang, X., Zhang, J.: Gait feature extraction and gait classification using two-branch CNN. Multimedia Tools Appl. **79**(3–4), 2917–2930 (2019). https://doi.org/10.1007/s11042-019-08509-w
97. Wang, Y., Chen, Z., Wu, Q.M.J., Rong, X.: Deep mutual learning network for gait recognition. Multimedia Tools Appl. **79**(31–32), 22653–22672 (2020). https://doi.org/10.1007/s11042-020-09003-4
98. Wang, Y., Song, C., Huang, Y., Wang, Z., Wang, L.: Learning view invariant gait features with two-stream GAN. Neurocomputing **339**, 245–254 (2019). https://doi.org/10.1016/j.neucom.2019.02.025
99. Wen, J., Wang, X.: Cross-view gait recognition based on residual long short-term memory. Multimedia Tools Appl. **80**(19), 28777–28788 (2021). https://doi.org/10.1007/s11042-021-11107-4
100. Wen, J., Wang, X.: Gait recognition based on sparse linear subspace. IET Image Process. **15**(12), 2761–2769 (2021). https://doi.org/10.1049/ipr2.12260
101. Wu, C., Song, Y., Zhang, Y.: Multi-view gait recognition using NMF and 2DLDA. Multimedia Tools Appl. **78**(24), 35789–35811 (2019). https://doi.org/10.1007/s11042-019-08153-4
102. Wu, H., Tian, J., Fu, Y., Li, B., Li, X.: Condition-aware comparison scheme for gait recognition. IEEE Trans. Image Process. **30**, 2734–2744 (2020). https://doi.org/10.1109/TIP.2020.3039888
103. Xu, K., Jiang, X., Sun, T.: Gait recognition based on local graphical skeleton descriptor with pairwise similarity network. IEEE Trans. Multimedia **24**, 3265–3275 (2021). https://doi.org/10.1109/TMM.2021.3095809
104. Xu, Z., Lu, W., Zhang, Q., Yeung, Y., Chen, X.: Gait recognition based on capsule network. J. Vis. Commun. Image Represent. **59**, 159–167 (2019). https://doi.org/10.1016/j.jvcir.2019.01.023
105. Xue, W., Ai, H., Sun, T., Song, C., Huang, Y., Wang, L.: Frame-GAN: increasing the frame rate of gait videos with generative adversarial networks. Neurocomputing **380**, 95–104 (2019). https://doi.org/10.1016/j.neucom.2019.11.015
106. Yang, Z., Zhong, X., Zhong, Z., Liu, H., Wang, Z., Satoh, S.: Win-win by competition: auxiliary-free cloth-changing person re-identification. IEEE Trans. Image Process. **32**, 2985–2999 (2023). https://doi.org/10.1109/TIP.2023.3277389

107. Yao, L., Kusakunniran, W., Zhang, P., Wu, Q., Zhang, J.: Improving disentangled representation learning for gait recognition using group supervision. IEEE Trans. Multimedia (2022). https://doi.org/10.1109/TMM.2022.3171961
108. Yeo, S.-S., Rho, S., Kim, H., Safdar, J., Zia, U., Durrani, M.Y.: A triplet-branch convolutional neural network for part-based gait recognition. Comput. Syst. Sci. Eng. **47**(2), 2027–2047 (2023). https://doi.org/10.32604/csse.2023.040327
109. Yusuf, S.I., Adeshina, S.A., Boukar, M.M.: Upper gait analysis for human identification using convolutional—recurrent neural network. J. Theor. Appl. Inform. Technol. **100**(13), 4968–4977 (2022). https://www.scopus.com/inward/record.uri?eid=2-s2.0-85134376226&partnerID=40&md5=07cbc793c37922db44879e02171c538f
110. Zell, P., Rosenhahn, B.: Learning inverse dynamics for human locomotion analysis. Neural Comput. Appl. **32**(15), 11729–11743 (2020). https://doi.org/10.1007/s00521-019-04658-z
111. Zhang, Y., Huang, Y., Yu, S., Wang, L.: Cross-view gait recognition by discriminative feature learning. IEEE Trans. Image Process. **29**, 1001–1015 (2019). https://doi.org/10.1109/TIP.2019.2926208
112. Zhang, Z., Tran, L., Liu, F., Liu, X.: On learning disentangled representations for gait recognition. IEEE Trans. Pattern Anal. Mach. Intell. **44**(1), 345–360 (2022). https://doi.org/10.1109/TPAMI.2020.2998790
113. Zhao, K., Zhang, M., Ji, J., Zhang, R., Bewley, J.M.: Automatic lameness scoring of dairy cows based on the analysis of head- and back-hoof linkage features using machine learning methods. Biosyst. Eng. **230**, 424–441 (2023). https://doi.org/10.1016/j.biosystemseng.2023.05.003
114. Zhao, N., Zhang, Z., Wang, Y., Wang, J., Li, B., Zhu, T., Xiang, Y.: See your mental state from your walk: Recognizing anxiety and depression through Kinect-recorded gait data. PLoS One **14**(5) (2019). https://doi.org/10.1371/journal.pone.0216591
115. Zheng, L., Zha, Y., Kong, D., Yang, H., Zhang, Y.: Multi-branch angle aware spatial temporal graph convolutional neural network for model-based gait recognition. IET Cyber-Syst. Robot. **4**(2), 97–106 (2022). https://doi.org/10.1049/csy2.12052
116. Zhu, H., Zhang, Y., Li, G., Zhang, J., Shan, H.: Ordinal distribution regression for gait-based age estimation. Sci. China Inform. Sci. **63**(2) (2020). https://doi.org/10.1007/s11432-019-2733-4
117. Zulcaffle, T.M.A., Kurugollu, F., Kipli, K., Joseph, A., Bong, D.B.L.: Front and back views gait recognitions using EfficientNets and EfficientNetV2 models based on gait energy image. Int. J. Comput. Digit. Syst. **14**(1), 749–758 (2023). https://doi.org/10.12785/ijcds/140157

Chapter 4
Datasets

Abstract This chapter delves into the datasets pivotal to gait identification research, offering a comprehensive analysis of those frequently employed in the reviewed articles. The chapter's focus is the CASIA-B dataset, prevalent in 51 out of 56 reviewed articles, highlighting its dominance. Other datasets such as OULP, OUMVLP, and TUM-GAID are also considered, each with unique attributes and applications. Through meticulous evaluation of these datasets, the chapter illuminates their respective performance metrics, facilitating nuanced comparisons across studies. By dissecting the datasets' nuances and comparing the results from different research works, this chapter offers a panoramic view of the dataset landscape in gait identification research.

4.1 Analysis of Popular Datasets in Gait Identification Research

In this section, the focus shifts to the datasets used by the reviewed articles. Each dataset is introduced, accompanied by accuracy results for the most frequently used ones, facilitating comparisons. This exploration aims to illuminate prevalent datasets in gait identification studies, providing insights into their respective performance metrics for comprehensive analysis and comparison.

In exploring gait identification studies, understanding the frequency with which various datasets are employed provides valuable insights into the landscape of research methodologies. The graphic of Fig. 4.1 delves into the recurring use of specific datasets across the studies included in the systematic review. Each dataset is represented by a distinct bar, allowing for a visual assessment of its prevalence in the broader context of gait identification research.

Out of 56 reviewed articles, 51 mainly feature the CASIA-B dataset [54], underscoring its clear popularity in gait identification research. The subsequent analysis will showcase accuracy results for the four most widely employed datasets, including CASIA-B. Analyzing the graphic in Fig. 4.1, the remaining datasets are: OULP [21], OUMVLP [37] and TUM-GAID [18]. This can also be seen in Table 4.1, which summarizes all the datasets mentioned in the articles included in this review.

D. R. M. Bastos and J. M. R. S. Tavares, *Advances in Gait-Based Identification*, Studies in Systems, Decision and Control 593,
https://doi.org/10.1007/978-3-031-89560-9_4

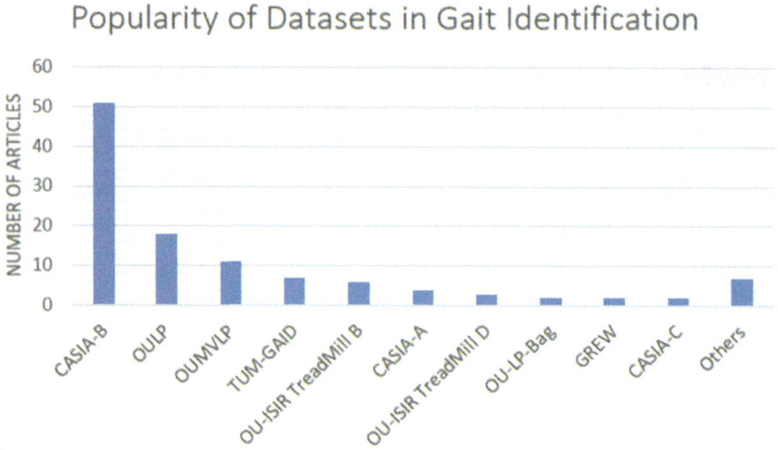

Fig. 4.1 Frequency of using datasets in the reviewed studies

4.2 CASIA Gait Database

The CASIA Gait Database, curated by the National Laboratory of Pattern Recognition (NLPR) at the Institute of Automation, Chinese Academy of Sciences (CASIA), comprises five datasets. These five datasets are CASIA-A [41], CASIA-B [54], CASIA-C [38], CASIA-D [58], and CASIA-E [36]. CASIA-A, CASIA-B, and CASIA-C are the datasets used for model training and evaluation in the reviewed articles. CASIA-D contains images related to plantar pressure, which falls outside the scope of this review. Despite its comprehensive nature, the CASIA-E dataset was not used in any reviewed articles. This absence is attributed to the dataset's recent release in 2023. Despite its non-utilization, the CASIA-E dataset is discussed due to its completeness and potential relevance to gait identification research.

4.2.1 CASIA-A

The CASIA-A dataset, a trailblazer in gait recognition studies since its inception in 2001, encompasses gait data from 20 individuals acquired in 12 image sequences from three distinct viewpoints: parallel, 45°, and 90° to the image plane as shown in Fig. 4.2. Each person contributes 4 sequences for each of these three directions. The sequence length varies to accommodate the range of walker speeds, spanning from 37 to 127 frames. The dataset comprises about 2.2 GB of data, incorporating a database of 19,139 images [41].

This dataset, initially instrumental in the early years of gait recognition research, is known for its straightforward design. However, with more complex datasets available

Table 4.1 Summary of the available datasets for gait identification in the reviewed articles

Dataset	#Subjects	#Sequences	#Views	Environment	Other factors	#Articles
CASIA-A	20	240	3	Static outdoor	None	4
CASIA-B	124	13.640	11	Static indoor	Bag carrying, dressing	51
CASIA-C	153	1.530	1	Static outdoor	Bag carrying, walking speed	2
CASIA-E	1.014	778.752	26	Multiple outdoor	Bag carrying, walking style, dressing, soft biometric features	0
TUM-GAID	305	3.370	1	Static indoor	Bag carrying, shoes	7
OU-ISIR A	34	612	1	Static indoor	Walking speed	1
OU-ISIR B	68	2.764	1	Static indoor	Dressing	6
OU-ISIR D	185	370	1	Static indoor	Gait fluctuations	3
OULP	4.007	7.842	2	Static indoor	None	18
GaitST	179	480	1	Static indoor	Acceleration, walking speed	1
OULP-Age	63.846	63.846	1	Static indoor	Age	1
OU-LP-Bag	62.528	187.584	1	Static indoor	Bag carrying	2
OU-LP-Bag β	2.068	4.136	1	Static indoor	Bag carrying	1
OUMVLP	10.307	288.596	14	Static indoor	None	11
OUMVLP-Pose	10.307	288.596	14	Static indoor	None	1
GREW	26.345	128.671	–	Wild	Diverse	2
DenseGait	217.954	217.954	–	Wild	Diverse	1
Gait3D	4.000	25.309	–	Wild	Diverse	1

and the preference for larger cohorts, CASIA-A's utility has diminished in recent years. The limited number of individuals in the dataset has contributed to this shift in preference.

Fig. 4.2 Visualization of three different points of view: 90°, parallel and 45°, respectively [41]

4.2.2 *CASIA-B*

The CASIA Dataset B, or CASIA-B, is the most commonly used dataset for evaluating performance in gait recognition algorithms, as seen in the graph in Fig. 4.1.

CASIA-B is a large dataset of gait videos, created in January 2005, consisting of gait data from 124 participants acquired from 11 different angles. For each participant, there are three variations, namely the normal gait (NM), clothing (CL), and transport element (BG), as shown in Fig. 4.3.

Six experimental tests were carried out for the normal case, two tests for the case where the person is wearing a coat, and two other tests for the case where the person is carrying an object.

In addition to video files, they also provide human silhouettes extracted from video files and GEI images. Gait silhouettes are the only publicly available data that can be obtained without resorting to any formal request or signing of an agreement.

Fig. 4.3 Video frames for a participant with **a** variation between the 11 capture angles and representation for each state, **b** normal (NM), **c** wearing a coat (CL), or **d** wearing a backpack (BG) [54]

The video name format is "xxx-mm-nn-ttt.avi" and the image name format is "xxx-mm-nn-ttt-yyy.png", where:

1. **xxx**: subject number. Ranges from 001 to 124;
2. **mm**: gait status, which can be "nm" (normal), "cl" (in a jacket), or "bg" (with an object);
3. **nn**: experimental test number. Can go from "01" to "06";
4. **ttt**: capture angle. 11 variations which are "000", "018", . . ., "180";
5. **yyy**: silhouette number for a specific angle of a specific test for a specific person. It can range from '001' to '999'.

In this dataset, the human silhouettes are 1,046,932 black and white images, where black corresponds to the background, and white corresponds to the participant's silhouette. These were acquired at a rate of 25 FPS and have a resolution of 240 × 320 pixels. The number of images per experimental test, per acquisition angle, varies according to the individual's pace, with an average of 77 frames per sequence.

To compile the accuracy values achieved by models in the articles reviewed for this dataset, it is essential to consider the metrics used for model evaluations. It is not equitable to compare the accuracy of a model tested under identical viewing angles between the gallery and the probe with the accuracy values of models tested under cross-view conditions. This refinement addresses a limitation in the review article Khaliluzzaman et al. [23], which inadvertently aggregates all accuracy values for NM, BG, and CL, regardless of the evaluation conditions. This approach can lead to misleading interpretations of the results.

Therefore, for concise and easily interpretable listing, a distinction is made between two evaluation conditions, identical-view and cross-view conditions, and two additional conditions: subjects divided or not divided. The models are evaluated using 4 normal sequences for training and 2 normal, 2 clothing, and 2 carrying sequences for testing in cases where the subjects are not divided. In most cases, the subjects are split into training and testing sets, with the testing set further divided into gallery and probe sets. In these cases, most approaches use the first 74 subjects for training and the remaining 50 for testing. For these 50 subjects, 4 normal sequences are used as the gallery, and the remaining 2 normal, 2 clothing, and 2 carrying sequences are used as the probe. In both cases, the results are divided into three categories. The notations "NM-NM", "NM-CL" and "NM-BG" are used to simplify these categories. The first "NM" corresponds to the gallery, representing the training set for undivided subjects. The "NM", "CL" and "BG" annotations following the "–" indicate the probe. In the training set, many studies use the first 4 normal sequences for training and the other two for testing, while a few use the first two sequences of each person for testing and the rest for training. This distinction was considered indifferent. The results are presented in Table 4.2.

The "✓" symbol in the "Cross-View Conditions" column indicates whether the results listed were obtained under cross-view conditions. For cross-condition, the only change is that angles identical to the test viewing angle were excluded. The highest values in each column are in bold for cross-view conditions and underlined

Table 4.2 Evaluation results of proposed models on the CASIA-A dataset under various test conditions, providing insights into performance across normal, clothing, and carrying sequences. Distinctions are made between "Identical-View Conditions" and "Cross-View Conditions". It also indicates if the subjects were divided or not

Name/References	Cross-view conditions	Subjects divided	NN-NN (%)	NN-BG (%)	NN-CL (%)	Average (%)
OPGEINet [17]		✓	97.80	45.3	49.6	64.23
LSTM-VGRNet2 [39]		✓	99.16	?	?	?
ACL + local + temporal [55]	✓	✓	96	?	?	?
CCGI + CNN [27]		✓	95.8	?	?	?
2D-CNN-SMP [4]			99.8	?	?	?
CCGI + SDCNN [10]		✓	98.87	85.83	91.01	91.90
Wang et al. [43]	✓	✓	94.5	88.5	66.9	83.30
GaitPart [7]	✓	✓	96.2	91.5	78.7	88.80
Moghaddam and Etemad [33]	✓	✓	95.2	89.7	74.7	86.53
GAEI [12]			98.78	90.24	66.66	85.23
Wu et al. [46]	✓	✓	94.8	88.8	81.8	88.47
GaitSet [5]	✓	✓	96.1	90.8	70.3	85.73
3D-CNN [11]		✓	98.77	?	?	?
DGEI + GAN [13]			100.00	96.37	68.14	88.17
LGSD + PSN [50]		✓	98.06	85.04	64.22	82.44
Gait-D [8]	✓	✓	91.6	79	72	80.87
CCGAN [1]			97.98	79	51.8	76.26
CNN-CapsNet [31]		✓	96.1	85.6	84.3	88.67
UGaitNet [30]	✓	✓	96.4	92.8	71.8	87.00
Han et al. [15]	✓	✓	96	91.6	74.8	87.47
GCAN [19]	✓	✓	98.51	95.37	84.51	92.80
Yao et al. [52]	✓	✓	92	85.9	71.5	83.13
MAST-GCN [57]	✓	✓	88.2	75.2	73	78.80
Gao et al. [9]	✓	✓	?	?	78.5	?
GaitGL [26]	✓	✓	98	95.4	**87.3**	**93.57**
ViTs16 [34]	✓	✓	**99.64**	79.51	24.45	67.87
Hua et al. [20]	✓	✓	96.5	92.6	79.8	89.63
IGPS [42]	✓	✓	98.6	**95.6**	85.1	93.10
Gait-CNN-ViT [32]		✓	100	?	?	?
Yeo et al. [53]			96.29	57.41	65.26	72.99
Pose-based BEI [14]			98.79	93.54	59.27	83.87

for identical-view conditions. The symbol "✓" in the "Subjects Divided" column indicates whether the results listed were obtained by dividing or not the subjects.

Other evaluation metrics were considered, however, those presented were the most common, allowing for a fair comparison. For example, evaluations such as "BG-BG" and "CL-CL" were also performed, mostly in years before 2020. In the years 2022 and 2023, no evaluations following these formats were found.

Analyzing the results reveals several insights. Firstly, there is a clear trend towards evaluating algorithms under cross-view conditions. The most cited and relevant articles in the field of gait recognition currently utilize these evaluation metrics. Interestingly, the highest average accuracy algorithm is GaitCL [26] with 93.57%, evaluated under cross-view conditions. This value surpasses the 91.90% achieved by CCGI + SDCNN [10], which should theoretically yield higher accuracy values when evaluated without excluding identical-view cases, but that is not the case. This highlights the outstanding performance of the GaitGL algorithm.

Examining the "NN-CL" case, the CCGI + SDCNN algorithm demonstrated the best performance, showcasing robustness under different clothing conditions. For the "NM-NM" case under cross-view conditions, the ViTs16 model [34] exhibited the highest performance with 99.64%. However, this algorithm proves to be sensitive under different clothing conditions, with an accuracy of 24.45% for "NN-CL". The DGEI + GAN algorithm [13], evaluated under identical-view conditions, achieved the highest accuracy for "NM-NM" and "NM-BG" with 100 and 96.37%, respectively. However, these results were obtained using all 124 people for training and testing without dividing the subjects, which leads to better results theoretically.

Yet under identical-view condition, but splitting the subjects, Gait-CNN-ViT [32] achieved the best result for "NM-NM" with 100%. However, this result was obtained using only 10% of the subjects for testing, corresponding to just 12 people. Achieving 100% accuracy is much easier with this smaller group than with the standard 50 people used for testing.

Among those evaluated under cross-view conditions, IGPS [42] demonstrated the best performance under different carrying conditions, with an accuracy of 95.6%.

4.2.3 CASIA-C

The CASIA-C dataset [38] was acquired using an infrared thermal camera during July and August 2005, featuring 153 subjects. The dataset encompasses four distinct walking conditions: normal walking (fn), slow walking (fs), fast walking (fq), and normal walking with a bag (fb). Each subject contributes 10 sequences (4 fn, 2 fs, 2 fq, and 2 fb), all recorded at night with a resolution of 320 × 240 as shown in Fig. 4.4. The sequences include variations in walking speeds and carrying conditions, adding diversity to the dataset. The evaluation process often involves identifying

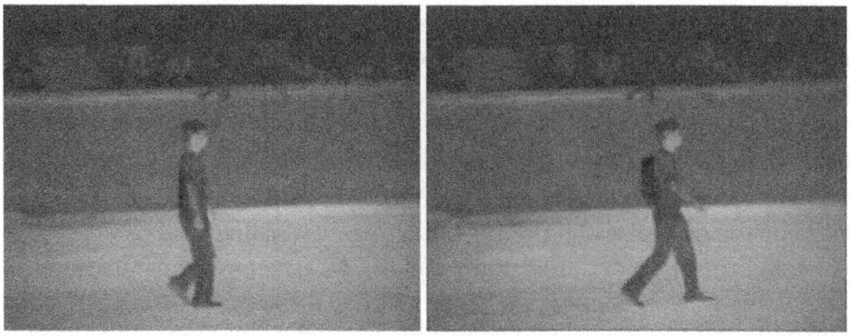

Fig. 4.4 Example of a frame acquired with an infrared thermal camera for a subject walking normally on the left and the same subject carrying a bag on the right [38]

individuals engaging in cross-speed walks, enhancing the dataset's applicability in gait recognition research.

4.2.4 CASIA-E

The CASIA-E dataset is the most recent and potentially the most comprehensive among the datasets covered in this state-of-the-art review. It boasts rich attributes, featuring a substantial number of subjects (1014), an extensive collection of gait videos, and variations in both spatial and temporal dimensions. With over a thousand individuals contributing to nearly one million videos, each person is captured from 13 view angles ranging from 0° to 180° at 15° intervals, as shown in Fig. 4.5. The dataset includes videos recorded with both low and high cameras for each angle, introducing also a diverse range of appearances influenced by factors such as bag carrying, dressing, and walking styles [36].

The dataset's richness extends to its temporal and environmental aspects. The videos were acquired over five months across three outdoor scenes, each presenting varying difficulty levels for the recognition task. Scenes include a simple background with a wall, a more complex setting, and a dynamic background featuring passing cars. Additionally, recordings in each scene were conducted at two different times of the day, morning and afternoon, to capture distinct lighting conditions.

Beyond gait-related information, the dataset records soft biometric features for all subjects, including age, gender, height, weight, and nationality. This multifaceted dataset provides a valuable resource for gait recognition research, offering a comprehensive and realistic representation of human gait in diverse real-world scenarios.

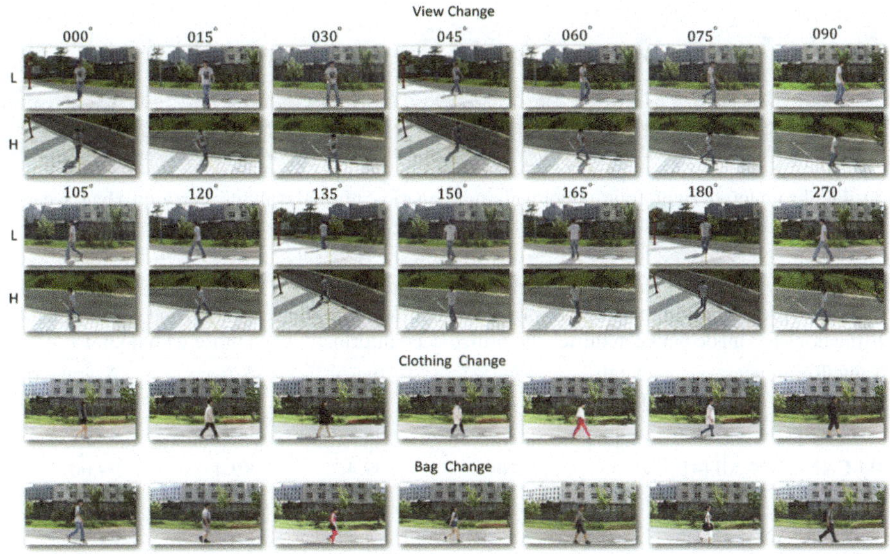

Fig. 4.5 Set of frames illustrating the change in the gait acquisition angle during normal walking, acquired by cameras positioned at both High (denoted as H) and Low (denoted as L) positions, featuring variations in clothing and bag [36]

4.3 TUM-GAID

The TUM-GAID dataset is a comprehensive collection that includes RGB video, depth, and audio data from 305 subjects engaged in walking under four conditions: normal walking (N), carrying a backpack (B), wearing coating shoes (S), and elapsed time (TN-TB-TS). Each subject contributes ten sequences, including six normal walking sequences (N1-N6), two backpack-carrying sequences (B1-B2), and two sequences with coating shoes (S1-S2). Additionally, 32 subjects have recordings in January and April, resulting in 10 extra sequences each (TN1-TN6, TB1-TB2, TS1-TS2). The dataset offers various scenarios, capturing variations in clothing, footwear, and lighting conditions. The Microsoft Kinect sensor records the action in a 640×480 pixels resolution with a frame rate of approximately 30 fps. Figure 4.6 covers left-to-right and right-to-left indoor walking trajectories. The comprehensive nature of TUM-GAID, with variations in carrying conditions and shoe types, makes it a valuable resource for gait recognition research [18].

It is important to note that articles using this dataset typically leverage the depth of information provided. This would be sufficient to exclude them from this review, considering the exclusion criteria outlined in Sect. 3.2. However, these documents are included because they also assess algorithm performance on other datasets that do not use depth information, such as the CASIA-B dataset. Therefore, this is briefly addressed by presenting the main results in Table 4.3, distinguishing between those using the depth information provided by the dataset and those that do not. Accuracy

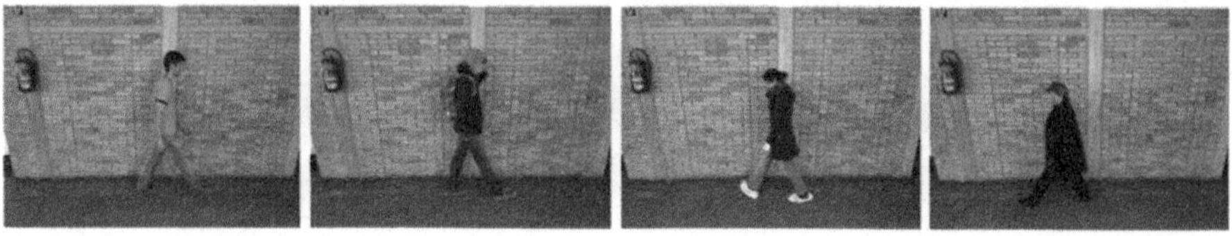

Fig. 4.6 From left to right, four indoor walking scenarios: normal walking, carrying a bag, wearing coated shoes, and wearing a coat [18]

Table 4.3 Evaluation results of proposed models on the TUM-GAID dataset under various testing conditions, providing insights into performance on different normal, clothing, and shoe sequences. Distinctions are made between those who utilize deep information and those who do not

Name/References	Depth information	N (%)	B(%)	S(%)	Average (%)
JITN [25]	✓	99.7	98.1	99.4	99.07
3D-CNN-7NN-All [4]	✓	**100**	99.4	99.4	99.60
GAEI [12]		76.48	<u>99.34</u>	<u>98.02</u>	<u>91.28</u>
DGEI + GAN [13]		94.18	73.81	86.54	84.84
UGaitNet [30]	✓	99.7	99.4	98.1	99.07
Han et al. [15]	✓	**100**	**100**	**99.7**	**99.90**
Pose-based BEI [14]		<u>94.76</u>	72.54	80.87	82.72

values were obtained considering the N1-N4 sets for training and the N5-N6, B1-B2, and S1-S2 sets for testing, represented in the table by "N", "B" and "S", respectively. None of the articles explicitly mention excluding identical viewpoints between the gallery and the probe.

In Table 4.3, the symbol "✓" in the "Depth Information" column indicates whether the results listed were obtained considering depth information. The highest values in each column are in bold for those using depth information and underlined for those without.

As expected, all articles using depth information achieve higher average accuracy values than those without. The article by Han et al. [15] stands out among those using depth information, boasting an average accuracy value of 99.90%. Among the three articles that do not use depth information and align with the criteria for this review, the GAEI algorithm [12] stands out with a significantly higher average accuracy than the second-best performer, presenting 91.28% accuracy. For this algorithm, the values for B and S even match those of algorithms using depth information. However, it shows the lowest accuracy for normal walking (N), with 76.48%.

4.4 OU-ISIR Gait Database

The OU-ISIR Gait Database was developed by the Institute of Information Systems and Industrial Research (ISIR) at Osaka University (OU), Japan. This database is widely recognized in gait identification research and provides diverse controlled conditions for capturing multi-person gait videos.

The OU-ISIR Gait Database encompasses 10 datasets, each designed to assess different aspects of gait identification. Among these, the following seven are elaborately discussed here:

1. Treadmill Database: This database comprises four datasets labeled as Dataset A, B, C, and D, each presenting a specific variation for study [28];
2. Large Population Dataset (OULP) [21];
3. Speed Transition Dataset (GaitST): This includes two datasets and a third to assist the training process [29];
4. Large Population Dataset with Age (OULP-Age) [49];
5. Large Population Dataset with Bag (OU-LP-Bag): Linked to this dataset is also a β version, which is here discussed shortly [40];
6. Multi-view Large Population Dataset (OUMVLP) [37];
7. Multi-view Large Population Dataset with Pose Sequence (OUMVLP-Pose) [2].

These datasets were chosen due to their relevance in the articles included in the systematic review. However, the remaining three, "Multi-view Large Population Dataset with Human Meshes," "Inertial Sensor Dataset," and "Similar Actions Inertial Dataset," are excluded from the more detailed analysis as they fall outside the scope of the review theme and were not referenced in any works included in the systematic review. Subsequently, each selected seven datasets is presented with more details to comprehensively understand their characteristics and contributions to gait identification research.

4.4.1 Treadmill Dataset

The OU-ISIR Treadmill Database, created at the Institute of Scientific and Industrial Research (ISIR), Osaka University (OU), comprises four distinct datasets, denoted as A, B, C, and D. These datasets were collected since March 2007, with the consent of the subjects. The data include individuals walking on a treadmill, acquired by 25 cameras at 60 fps with a 640×480 pixels resolution. The datasets are distributed in silhouette sequences size-normalized to 88×128 pixels [28]. As Dataset C is currently under preparation, it is not discussed here.

1. Treadmill Dataset A—Speed Variation: The dataset A consists of gait silhouette sequences of 34 subjects, focusing on speed variation from 2 to 10 km/h at 1 km/h intervals. The subjects walked and ran, contributing to the study of gait at various speeds;

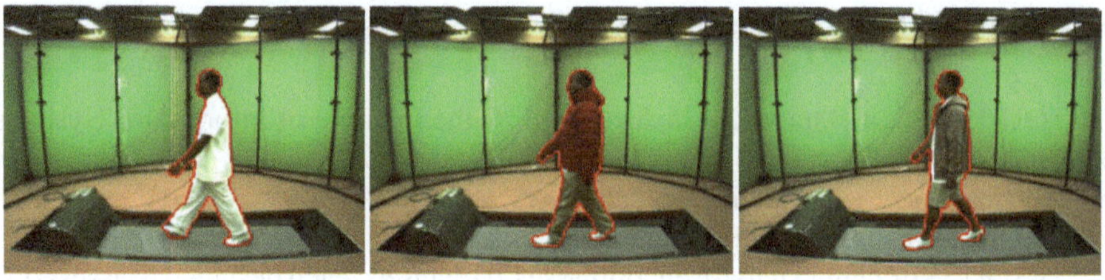

Fig. 4.7 Three examples of different types of clothing in the OU-ISIR Treadmill B dataset [28]

2. Treadmill Dataset B—Clothes Variation: Treadmill Dataset B is an extensive collection of gait silhouette sequences from the front-parallel view, featuring 68 subjects and focusing on variations in clothing. This dataset stands out as one of the most comprehensive resources for studying the impact of clothing diversity on gait recognition algorithms. With up to 32 combinations of clothing types, including variations such as half shirts, medical doctor pants, down jackets, baggy pants, parkas, half pants, and more, Dataset B offers a rich and diverse set of conditions. Figure 4.7 presents some clothing variations examples. The subjects' gait sequences, recorded twice on the same day, thoroughly explore the challenges of different clothing variations. The dataset is divided into three parts: training set, gallery set, and probe set. The training set consists of data from 20 subjects exhibiting variations in clothing ranging from 15 to 28 combinations. The gallery set comprises the remaining 48 subjects, each acquired in a standard type of clothing. On the other hand, the probe set includes all sequences from these 48 subjects but without the standard type of clothing, ensuring a comprehensive evaluation of gait recognition algorithms under various clothing conditions. Researchers commonly rely on this dataset for both training and evaluating models, making it a pivotal resource in the domain of gait recognition;
3. Treadmill Dataset D—Gait Fluctuation: Dataset D encompasses gait silhouette sequences of 185 subjects, focusing on various gait fluctuations. The dataset is divided into two subsets, DB_{low} and DB_{high}, representing individuals with large and small gait fluctuations, respectively. This variation provides insights into understanding the influence of gait fluctuations on recognition algorithms.

Dataset B is the most extensively used among the four datasets, emerging prominently in the reviewed articles as a primary choice for training and model evaluation.

4.4.2 OULP

The OU-ISIR Large Population (OULP) dataset [21], established by the Institute of Scientific and Industrial Research (ISIR) at Osaka University, is widely known in the field of gait recognition for its extensive subject coverage. Collected since March 2009 through outreach activities in Japan, this dataset comprises gait images

Fig. 4.8 Gait frames of different people with different ages present in the OULP dataset [21]

from 4007 individuals, with a notable age range spanning from 1 to 94 years. This age range is shown in Fig. 4.8. The data collection adheres to ethical standards, obtaining informed consent from all participants. The images, distributed in silhouette sequences, are registered and size-normalized to dimensions of 88 × 128 pixels, ensuring consistency in the dataset.

This dataset comprises two main subsets: subset A for evaluation and subset B for training. Subset A encompasses 1912 subjects, each contributing two sequences, and is recorded from four different views (55°, 65°, 75°, and 85°), providing diverse perspectives due to varying camera angles. In contrast, subset B contains only one sequence per subject, likewise captured from the four different views.

Various experimental settings have been employed to evaluate models, emphasizing the need for careful consideration and comparison of articles operating under similar evaluation conditions. Table 4.4 lists articles that commonly adopted specific evaluation conditions. Within this table, some articles are associated with two different evaluation settings. Articles marked with the symbol "✓" in the "Subjects Splitted" column indicate that they divided the 1912 subjects in subset A equally, resulting in 956 subjects for both probe and gallery sets. This translates to two gait sequences per subject in the probe and gallery sets for each of the four views. Articles lacking the "✓" symbol in the "Subjects Splitted" column split the two sequences of each 1912 subjects, resulting in 1912 subjects for both gallery and probe sets. Each gallery and probe set contains one sequence per subject for the four different views.

While there are other evaluation settings used, such as in the article by Zhang et al. [55], which utilizes 3.075 unique subjects for training and 768 subjects for testing, these settings were not as prevalent in the documents reviewed for this study and, therefore, it is not further detailed.

The angle values in the header of Table 4.4 indicate the probe angle. The accuracy values for each probe angle represent the average of three accuracies obtained by evaluating that specific probe angle with each gallery angle. This evaluation excludes identical angles between the probe and gallery to simulate cross-view conditions. The bold values represent the highest among those that split subjects in the evaluation settings. The underlined values indicate the highest for those using the other evaluation metric.

Analyzing the accuracy values across different articles for the OULP dataset, two standout articles are Wu et al. [46] and Sayeed et al. [35], as they exhibit the highest accuracy values across all angles for evaluation settings where subjects are

Table 4.4 Evaluation results of proposed models on the OULP dataset under four different angles in two different evaluation settings

Name/References	Subjects splitted	55 (%)	65 (%)	75 (%)	85 (%)	Average (%)
TS-GAN [44]	✓	84.9	89.43	88.83	85	87.04
Wu et al. [45]		84.43	95.23	95.07	87.33	90.52
Frame-GAN [51]	✓	87.3	92.8	92.5	89.1	90.43
Wang et al. [43]		97.47	98.27	98.10	97.73	97.89
Wu et al. [46]		99.6	99.97	99.83	99.7	99.78
Sayeed et al. [35]	✓	**91.8**	**93.47**	**93.93**	**91.5**	**92.68**

not divided and are divided, respectively. As could be observed, there is a tendency for articles that divide subjects into gallery and probe sets to yield lower accuracy values. Splitting the probe and gallery sets among different subjects may represent real-world applications, where the model must recognize an unfamiliar person based on a single observation instance, reflecting a scenario where the model encounters an unknown individual and needs to identify them based on a previously seen set of people. Conversely, splitting probe and gallery among sequences of the same subjects can be more challenging, requiring the model to generalize beyond intra-subject variations. This simulates a situation where the model encounters a person wearing different clothes, carrying a bag, or displaying other variations in appearance and posture. In this case, as both sequences for each subject are recorded under the same conditions, the identification task becomes easier, resulting in higher accuracy values.

4.4.3 GaitST

The OU-ISIR Gait Speed Transition Database, also known as GaitST [29], is a valuable resource for evaluating gait identification algorithms, particularly under varying walking speeds. The database comprises two main datasets: Dataset 1 and Dataset 2.

Dataset 1 includes probe sets with speed-transitioned gait sequences recorded from 26 subjects in an indoor environment. Subjects were instructed to walk towards a poster, gradually decreasing their speed and coming to a stop. The probe set captures significant changes in stride during these transitions. The gallery set for Dataset 1 contains gait sequences from 179 subjects, including the 26 probe subjects. These subjects walked at a constant speed (4 km/h on a treadmill) or nearly constant speed (on the ground) for a few seconds.

Dataset 2 features probe sets with gait sequences from 25 subjects, each walking on a speed-controlled treadmill, undergoing accelerations from 1 to 5 km/h and decelerations from 5 to 1 km/h. The gallery set for Dataset 2 comprises gait sequences from 154 subjects, including the 25 probe subjects. Like Dataset 1, these subjects walked at a constant speed (4 km/h) for 6 s.

Additionally, the dataset includes an auxiliary training set with 24 subjects, each walking at speeds of 2, 3, 4, and 5 km/h. The parameters for gait recognition models were learned using this auxiliary training set, which contains images of resolution 32×22 pixels. All image sequences were captured at 60 fps, binarized through background subtraction-based graph-cut segmentation, and size-normalized to 22×32 pixels. The dataset offers a unique opportunity to explore gait recognition across varying speeds, making it an essential algorithm development and testing benchmark.

4.4.4 OULP-Age

O Large Population Dataset with Age (OULP-Age) [49] from the OU-ISIR Gait Database is designed to facilitate research in the development, testing, and evaluation of algorithms for human gait analysis, particularly about age and gender factors, such as gait-based age estimation.

The data collection process was aligned with a long-run exhibition of video-based gait analysis at a science museum, where explicit informed consent was obtained from all participating subjects. The dataset comprises video frames capturing the walking activities of an extensive population, totaling 63,846 subjects, navigating a designated course recorded by a camera at 30 frames per second, with a resolution of 640×480 pixels. Notably, subjects in this dataset span a wide age range, from 2 to 90 years old.

A size-normalized GEI is provided for each subject, with dimensions of 88×128 pixels. Some examples of these GEIs are displayed in Fig. 4.9. The dataset also includes comprehensive information, such as subject ID lists for the training and testing sets, along with the age and gender details for each individual. This dataset aims to contribute to advancements in gait-based applications, offering valuable resources for researchers and practitioners interested in gait analysis, especially in age-related factors.

4.4.5 OU-LP-Bag

The OU-ISIR Large Population Gait Database with a real-life carried object (OU-LP-Bag) is globally accepted as the largest repository for gait analysis, boasting video frames capturing the walking activities of 62,528 diverse individuals in various scenarios. This database introduces a nuanced perspective by annotating seven distinctive carrying status labels: no carried objects (NoCO), objects on the side bottom (SbCO), side middle (SmCO), front (FrCO), back (BaCO), multiple objects

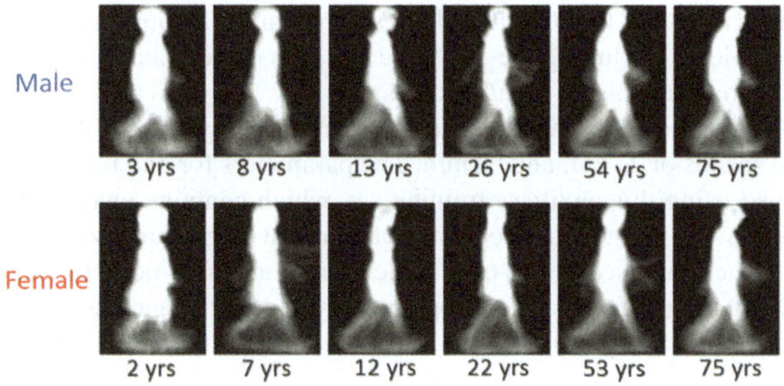

Fig. 4.9 A few GEIs depicting subjects of different ages from the OULP-Age dataset [49]

in different locations (MuCO), and changing carrying locations during walking (CpCO). Figure 4.10 shows examples of these labels. Subjects, ranging in age from 2 to 95 years, contribute three walking sequences each: two without carried objects and one featuring objects. The database was meticulously collected in conjunction with a science museum's video-based gait analysis demonstration, employing a camera positioned approximately 8 m from the walking course at a height of 5 m, with an image resolution of 1280×980 pixels and a frame rate of 25 fps. The first sequence (A1) for each subject may or may not include carried objects, while the subsequent sequences (A2 and A3) are without carried objects [40].

Previous to this database, there is also a β version (OU-LP-Bag β) that includes 2.068 subjects, each contributing two sequences: one with carried objects and the other without. This beta version is divided into three datasets: a training set with 2.068 sequences from 1.034 subjects and disjoint gallery and probe sets with 1.036 subjects each. The gallery set contains sequences without carried objects, while the probe set features sequences with carried objects [48].

These datasets are integral to advancing research in vision-based gait recognition, particularly when considering the covariate of carried objects. They include silhouette sequences and size-normalized (88×128 pixels) GEIs, Fig. 4.10, providing a valuable resource for algorithm development, testing, and evaluation, shedding light on real-life gait scenarios with carried objects.

4.4.6 OUMVLP

Created by the Institute of Scientific and Industrial Research (ISIR) at Osaka University (OU), the OU-ISIR Gait Database, Multi-View Large Population Dataset (OU-MVLP) [37], is a substantial dataset designed to support research endeavors in cross-view gait recognition algorithm development, testing, and evaluation. The data

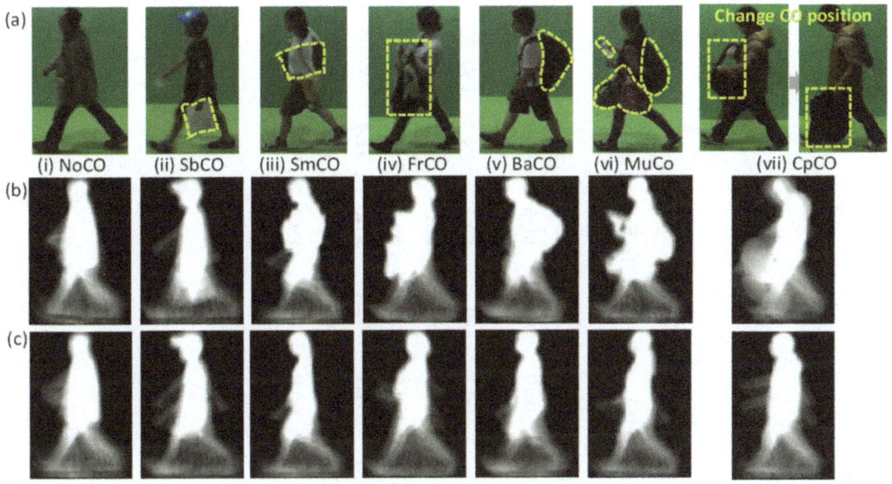

Fig. 4.10 Data provided by the OU-LP-Bag with **a** examples of seven frames with distinctive carrying status labels, **b** the respective GEIs, and **c** the same GEIs without the transport element [40]

collection took place during an extensive exhibition of video-based gait analysis at a science museum, where explicit informed consent was obtained from all participants.

Comprising a remarkable number of subjects (10,307), evenly distributed between genders (5114 males and 5193 females), and spanning ages from 2 to 87 years, the dataset captures gait images from 14 different view angles. These angles range from 0° to 90° and 180° to 270°, with a 15° angle change in each step, effectively covering a quarter of a circle. The gait images, acquired at 25 frames per second, boast a 1280×980 pixels resolution.

The dataset is divided into two subsets for research purposes: a training set and a testing set. Both subsets are nearly equal in size, providing a robust foundation for algorithm training and evaluation. The training set comprises 5153 individuals, while the testing set includes the remaining 5154 subjects. Each subject contributes data from 14 views, with each view containing two sequences. During the testing phase, sequences with index 01 form the galleries, while those with index 02 serve as probes. This evaluation protocol was followed by the reviewed articles in Table 4.5. Accuracy values are reported for each of the 14 angles in the probe and evaluated against all angles in the gallery, excluding identical-view cases.

Analyzing the accuracy values in Table 4.5, one can identify GaitGL [26] as the top-performing algorithm, achieving the highest accuracy for 13 out of the 14 gait acquisition angles. Examining angle by angle, a remarkable result is observed with 91.1% accuracy for the 0° angle, surpassing the second-best by 5.9%. This outcome is surprising given that many studies consider this angle to provide less discriminative information about gait. GaitGL's exceptional performance is reflected in its average accuracy of 92.04%, which is 2.23% higher than the second-best, IGPS [42]. It is

Table 4.5 Evaluation results of proposed models on the OUMVLP dataset under 14 different angles, in cross-view conditions. (The highest accuracy value for each angle is highlighted in bold.)

Name/References	0°	15°	30°	45°	60°	75°	90°	180°	195°	210°	225°	240°	255°	270°	Ave
ACL + local + temporal [55]	74	88.3	94.6	95.4	88	91.3	90	76.7	89.5	95	94.9	88	90.8	89.8	89.02
GaitPart [7]	82.6	88.9	90.8	91	89.7	89.9	89.5	85.2	88.1	90	90.1	89	89.1	88.2	88.72
GaitSet [5]	81.3	88.6	90.2	90.7	88.6	89.1	88.3	83.1	87.7	89.4	89.7	87.8	88.3	86.9	87.84
GCAN [19]	84.99	90.34	91.26	91.4	90.63	90.57	90.14	87.09	89.37	90.46	90.64	90.02	89.81	89.1	89.7
GaitGL [26]	**91.1**	**92.6**	**92.3**	**92.5**	**92.8**	**92.2**	**92.1**	**92.4**	**91.9**	91.7	**91.9**	**92.1**	**91.5**	**91.4**	**92.04**
ViTs16 [34]	78	88.06	91.11	90.69	86.8	87.69	85.86	83.2	90.57	**92.24**	91.88	87.1	88.32	86.36	87.71
IGPS [42]	85.2	90.5	91.3	91.5	90.6	90.7	90.3	87	89.8	90.6	90.9	89.8	89.9	89.3	89.81

important to note that the ViTs16 algorithm [34] achieved the highest accuracy for the 210° angle. However, it could not maintain the same level of performance across all angle values, presenting, for example, 78% accuracy for the 0° angle.

Other evaluation settings have been employed in the literature, but as they were not the majority, they are not considered in this analysis. An example of such a case is the study Moghaddam and Etemad [33] that only used 4 out of the 14 angles for conducting the tests.

4.4.7 *OUMVLP-Pose*

The Multi-view Large Population Database with Pose Sequence (OUMVLP-Pose) [2] is an integral component of the OU-ISIR Gait Database, designed to facilitate advancements in model-based gait recognition algorithms. Created by the Institute of Scientific and Industrial Research (ISIR) at Osaka University (OU), this dataset builds upon the OUMVLP and consists of 10,307 subjects engaged in round-trip walking sequences. The data acquisition was conducted using seven network cameras at 15° intervals, resulting in 14 distinct views per subject as they traversed the walking course. The images have a resolution of 1280×980 pixels, recorded at a frame rate of 25 fps.

Each gait sequence within OUMVLP-Pose encompasses an average of 25 frames, comprehensively representing individuals' walking patterns. Two state-of-the-art pose estimation algorithms, OpenPose [3] and AlphaPose [24], were employed to extract pose sequences from RGB images. These pose sequences include information about 18 joints, enhancing the granularity of the dataset. Some examples of estimated pose results using OpenPose, obtained from frames at a 45° angle, are depicted in Fig. 4.11.

The dataset is divided into two subsets for practical research applications: a training set and a testing set, both meticulously curated to maintain a balanced distribution of subjects. Importantly, the training and testing sets, derived from both OpenPose

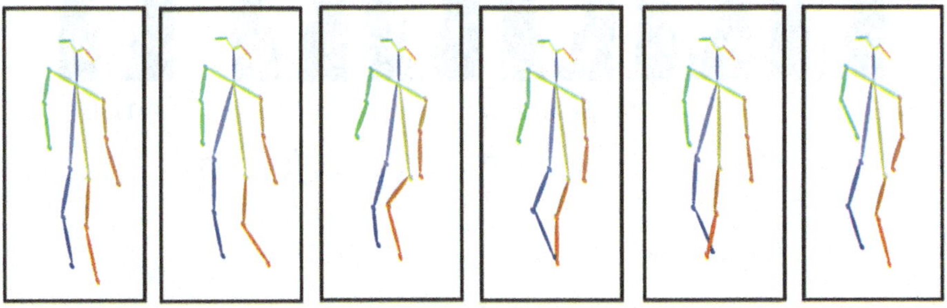

Fig. 4.11 Estimated pose results, using OpenPose, in frames at a 45° angle [2]

and AlphaPose datasets, maintain parity regarding the number of subjects and key parameters. This approach ensures a fair and consistent evaluation of algorithms.

OUMVLP-Pose is a valuable resource for researchers developing, testing, and evaluating gait recognition model-based algorithms. Its expansive coverage, incorporation of pose sequences, and compatibility with two leading pose estimation algorithms make it a robust dataset for advancing the state of the art in gait recognition research.

4.5 GREW

The GREW dataset [47] stands out as a comprehensive and large-scale benchmark designed for advancing research in unconstrained gait recognition. Comprising 26,345 subjects and 128,671 sequences, this "in-the-wild" dataset captures individuals across multiple days and diverse clothing styles, adding complexity to the recognition task. The dataset incorporates four distinct data types: Silhouettes, Optical Flow, 2D Pose, and 3D Pose, providing a rich set of information for both appearance-based and model-based algorithms. The 2D and 3D human pose estimation were obtained using algorithms presented in the studies [16, 22], respectively. These 2D and 3D key points are present in Fig. 4.12, along with images of silhouettes and their respective GEI.

The raw data, acquired from 882 cameras in open environments, introduces challenges such as occlusion, truncation, and varying illumination, setting it apart from conventional datasets like CASIA-B and OUMVLP.

In its construction, GREW leverages thousands of hours of streams from multiple cameras, resulting in over 14 million annotated boxes that offer silhouettes and human poses. Additionally, GREW introduces a novel element to the gait research community: a distractor set with 233,857 sequences. This set, comprising unlabeled data, enhances the dataset's suitability for real-world applications. Notably, GREW's

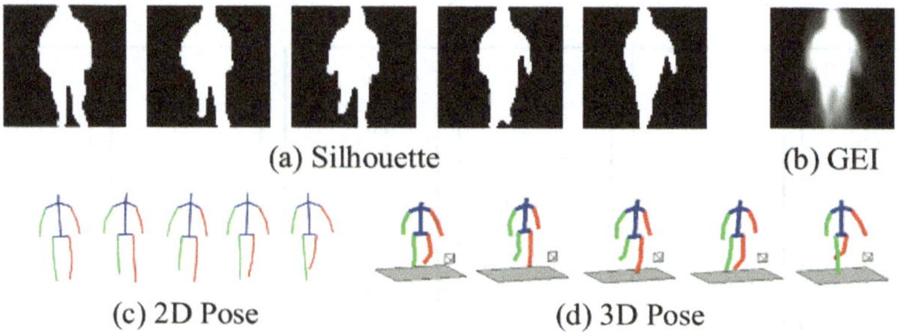

Fig. 4.12 Silhouette images provided by the GREW dataset and their corresponding GEI and 2D/3D key points for evaluating model-based algorithms [47]

meticulous annotation process and its provision of diverse attributes make it the first dataset for unconstrained gait recognition.

For training purposes, GREW is partitioned into three sets: the training set with 20,000 identities and 102,887 sequences, the validation set with 345 identities and 1784 sequences, and the test set with 6000 identities, each having four sequences. The dataset's size, diversity, and inclusion of a distractor set make it a valuable resource for advancing the state-of-the-art in gait recognition research.

4.6 DenseGait

DenseGait [6] stands as a substantial "in the wild" gait dataset designed for pre-training gait analysis systems, boasting a remarkable scale with 217,000 anonymized tracklets. The dataset was meticulously built by automatically processing surveillance video streams using AlphaPose, resulting in a diverse collection of gait sequences captured in various environments and camera angles across multiple geographical locations. The annotations within DenseGait primarily rely on tracking identifiers, introducing some noise due to automatic collection. While there might be instances of sequences belonging to the same subject, a rarity, this aligns with the characteristics of many large-scale unlabeled datasets with samples from the same semantic class, often considered unrelated during training.

DenseGait's uniqueness lies in its comprehensive annotation, featuring 42 appearance attributes for each of the 217,000 tracklets. An ensemble of pedestrian attribute identification networks contributed to the dense annotation, covering attributes such as gender, age group, body fat, camera viewpoint, clothing information, and apparent action. This dataset, focused on pre-training, is the largest gait dataset for such purposes. Unlike datasets meant for evaluation, DenseGait's automatic annotations render it unsuitable for ground-truth label assessments.

The dataset embraces real-world scenarios, showcasing walking individuals in natural settings and avoiding the artificial constraints imposed by laboratory conditions. DenseGait's markerless, non-treadmill nature encapsulates the full spectrum of factors influencing real-world gait patterns. Maintaining a commitment to privacy, DenseGait has been fully anonymized, eliminating any details that could lead to individual identification, including temporal, locational, and source information. DenseGait's availability to the research community adds a significant asset for advancing the field of gait analysis.

4.7 Gait3D

The Gait3D dataset [56] is a significant resource for researchers and practitioners in gait recognition. It comprises a vast collection of 3D representations of human gait, offering unprecedented insights into this biometric modality. With over 25.000

sequences derived from 4000 subjects, Gait3D stands out as one of the largest datasets of its kind. These sequences were meticulously acquired in a real-world setting, specifically within the confines of a large supermarket, providing a diverse and representative sample of human gait behaviors.

The dataset is partitioned into training and test sets to facilitate algorithm evaluation and development. The training set encompasses 3000 subjects, while the test set comprises the remaining 1000 subjects. Notably, each subject in the test set is represented by a single sequence designated as the probe set, with the remaining sequences forming the gallery set. This setup enables rigorous assessment of algorithm performance under realistic conditions.

Moreover, Gait3D offers a unique advantage over existing datasets by providing precise 3D human meshes recovered from video frames. These meshes offer detailed information about the pose and shape of human bodies and accurate viewpoint parameters. Additionally, the dataset includes conventional 2D silhouettes and key points, facilitating the exploration of gait recognition using multi-modal data. With its scalability, rich data modalities, and real-world relevance, Gait3D is a valuable resource for advancing research and applications in gait recognition.

References

1. Alsaggaf, W.A., Mehmood, I., Khairullah, E.F., Alhuraiji, S., Sabir, M., Alghamdi, A.S., Abd El-Latif, A.A.: A smart surveillance system for uncooperative gait recognition using cycle consistent generative adversarial networks (CCGANs). Comput. Intell. Neurosci. (2021). https://doi.org/10.1155/2021/3110416
2. An, W., Yu, S., Makihara, Y., Wu, X., Xu, C., Yu, Y., Liao, R., Yagi, Y.: Performance evaluation of model-based gait on multi-view very large population database with pose sequences. IEEE Trans. Biometrics Behav. Identity Sci. 1–1 (2020). https://doi.org/10.1109/TBIOM.2020.3008862
3. Cao, Z., Hidalgo, G., Simon, T., Wei, S.-E., Sheikh, Y.: Openpose: realtime multi-person 2d pose estimation using part affinity fields (2019)
4. Castro, F., Marín-Jiménez, M., Guil, N., Blanca, N.: Multimodal feature fusion for CNN-based gait recognition: an empirical comparison. Neural Comput. Appl. **32**(17), 14173–14193 (2020). https://doi.org/10.1007/s00521-020-04811-z
5. Chao, H., Wang, K., He, Y., Zhang, J., Feng, J.: GaitSet: cross-view gait recognition through utilizing gait as a deep set. IEEE Trans. Pattern Anal. Mach. Intell. **44**(7), 3467–3478 (2021). https://doi.org/10.1109/TPAMI.2021.3057879
6. Cosma, A., Radoi, E.: Learning gait representations with noisy multi-task learning. Sensors **22**(18) (2022). https://www.mdpi.com/1424-8220/22/18/6803
7. Fan, C., Peng, Y., Cao, C., Liu, X., Hou, S., Chi, J., Huang, Y., Li, Q., He, Z.: Gaitpart: temporal part-based model for gait recognition. In: IEEE/CVF Conference on Computer Vision and Pattern Recognition (CVPR), vol. 2020, pp. 14213–14221 (2020). https://doi.org/10.1109/CVPR42600.2020.01423
8. Gao, S., Yun, J., Zhao, Y., Liu, L.: Gait-D: skeleton-based gait feature decomposition for gait recognition. IET Comput. Vis. **16**(2), 111–125 (2021). https://doi.org/10.1049/cvi2.12070
9. Gao, Z., Wu, J., Wu, T., Huang, R., Zhang, A., Zhao, J.: Robust clothing-independent gait recognition using hybrid part-based gait features. PeerJ Comput. Sci. **8** (2022). https://doi.org/10.7717/peerj-cs.996

10. George, M., Govindarajan, T., Rajasekaran, K., Bandi, S.: A robust similarity based deep Siamese convolutional neural network for gait recognition across views. Comput. Intell. **36**(3), 1290–1319 (2020). https://doi.org/10.1111/coin.12361
11. Gul, S., Malik, M.I., Khan, G.M., Shafait, F.: Multi-view gait recognition system using spatio-temporal features and deep learning. Expert Syst. Appl. **179** (2021). https://doi.org/10.1016/j.eswa.2021.115057
12. Gupta, S., Chattopadhyay, P.: Exploiting pose dynamics for human recognition from their gait signatures. Multimedia Tools Appl. **80**(28–29), 35903–35921 (2020). https://doi.org/10.1007/s11042-020-10071-9
13. Gupta, S., Chattopadhyay, P.: Gait recognition in the presence of co-variate conditions. Neurocomputing **454**, 76–87 (2021). https://doi.org/10.1016/j.neucom.2021.04.113
14. Gupta, S., Chattopadhyay, P.: Pose-based boundary energy image for gait recognition from silhouette contours. Sadhana—Acad. Proc. Eng. Sci. **48**(4) (2023). https://doi.org/10.1007/s12046-023-02234-9
15. Han, F., Li, X., Zhao, J., Shen, F.: A unified perspective of classification-based loss and distance-based loss for cross-view gait recognition. Pattern Recogn. **125** (2022). https://doi.org/10.1016/j.patcog.2021.108519
16. Hang, C., Deva, R.: 3d human pose estimation = 2d pose estimation + matching (2017)
17. Hawas, A., El-Khobby, H., Elnaby, M., El-Samie, F.: Gait identification by convolutional neural networks and optical flow. Multimedia Tools Appl. **78**(18), 25873–25888 (2019). https://doi.org/10.1007/s11042-019-7638-9
18. Hofmann, M., Geiger, J., Bachmann, S., Schuller, B., Rigoll, G.: The tum gait from audio, image and depth (GAID) database: multimodal recognition of subjects and traits [Visual understanding and applications with RGB-D cameras]. J. Vis. Commun. Image Represent. **25**(1), 195–206 (2014). https://doi.org/10.1016/j.jvcir.2013.02.006
19. Hou, S., Liu, X., Cao, C., Huang, Y.: Gait quality aware network: toward the interpretability of silhouette-based gait recognition. IEEE Trans. Neural Netw. Learn. Syst. **34**(11), 8978–8988 (2022). https://doi.org/10.1109/TNNLS.2022.3154723
20. Hua, C., Pan, Y., Li, J., Wang, Z.:. Gait recognition by combining the long-short-term attention network and personal physiological features. Sensors **22**(22) (2022). https://doi.org/10.3390/s22228779
21. Iwama, H., Okumura, M., Makihara, Y., Yagi, Y.: The OU-ISIR gait database comprising the large population dataset and performance evaluation of gait recognition. IEEE Trans. Inform. Forensics Secur. **7**(5), 1511–1521 (2012). https://doi.org/10.1109/TIFS.2012.2204253
22. Ke, S., Bin, X., Dong, L., Jingdong, W.: Deep high-resolution representation learning for human pose estimation (2019)
23. Khaliluzzaman, M., Uddin, A., Deb, K., Hasan, M.J.: Person recognition based on deep gait: a survey. Sensors **23**(10) (2023). https://doi.org/10.3390/s23104875
24. Li, J., Wang, C., Zhu, H., Mao, Y., Fang, H.-S., Lu, C.: CrowdPose: efficient crowded scenes pose estimation and a new benchmark (2019)
25. Li, X., Makihara, Y., Xu, C., Yagi, Y., Ren, M.: Joint intensity transformer network for gait recognition robust against clothing and carrying status. IEEE Trans. Inform. Forensics Secur. **14**(12), 3102–3115 (2019). https://doi.org/10.1109/TIFS.2019.2912577
26. Lin, B., Zhang, S., Wang, M., Li, L., Yu, X.: GaitGL: learning discriminative global-local feature representations for gait recognition (2022)
27. Linda, G.M., Themozhi, G., Bandi, S.R.: Color-mapped contour gait image for cross-view gait recognition using deep convolutional neural network. Int. J. Wavelets Multiresolut. Inform. Process. **18**(1) (2019). https://doi.org/10.1142/S0219691319410121
28. Makihara, Y., Mannami, H., Tsuji, A., Hossain, M.A., Sugiura, K., Mori, A., Yagi, Y.: The OU-ISIR gait database comprising the treadmill dataset. IPSJ Trans. Comput. Vis. Appl. **4** (2012). https://doi.org/10.2197/ipsjtcva.4.53
29. Mansur, A., Makihara, Y., Aqmar, R., Yagi, Y.: Gait recognition under speed transition. In: IEEE Conference on Computer Vision and Pattern Recognition, vol. 2014, pp. 2521–2528 (2014). https://doi.org/10.1109/CVPR.2014.323

30. Marin-Jimenez, M.J., Castro, F.M., Delgado-Escano, R., Kalogeiton, V., Guil, N.: UGaitNet: multimodal gait recognition with missing input modalities. IEEE Trans. Inform. Forensics Secur. **16**, 5452–5462 (2021). https://doi.org/10.1109/TIFS.2021.3132579
31. Merlin Linda, G., Lakshmi, Sree Rathna, N.V.S., Murugan, N.S., Mahapatra, R.P., Muthukumaran, V., Sivaram, M.: Intelligent recognition system for viewpoint variations on gait and speech using CNN-CapsNet. Int. J. Intell. Comput. Cybern. **15**(3), 363–382 (2021). https://doi.org/10.1108/IJICC-08-2021-0178
32. Mogan, J., Lee, C., Lim, K., Ali, M., Alqahtani, A.: Gait-CNN-ViT: multi-model gait recognition with convolutional neural networks and vision transformer. Sensors **23**(8) (2023). https://doi.org/10.3390/s23083809
33. Moghaddam, A., Etemad, A.: View-invariant gait recognition with attentive recurrent learning of partial representations. IEEE Trans. Biometrics Behav. Identity Sci. **3**(1), 124–137 (2020). https://doi.org/10.1109/TBIOM.2020.3031470
34. Pinčić, D., Sušanj, D., Lenac, K.:. Gait recognition with self-supervised learning of gait features based on vision transformers. Sensors **22**(19) (2022). https://doi.org/10.3390/s22197140
35. Sayeed, M.S., Min, P.P., Bari, M.A.: Deep learning based gait recognition using convolutional neural network in the COVID-19 pandemic. Emerg. Sci. J. **6**(5), 1086–1099 (2022). https://doi.org/10.28991/ESJ-2022-06-05-012
36. Song, C., Huang, Y., Wang, W., Wang, L.: CASIA-E: a large comprehensive dataset for gait recognition. IEEE Trans. Pattern Anal. Mach. Intell. **45**(3), 2801–2815 (2023). https://doi.org/10.1109/TPAMI.2022.3183288
37. Takemura, N., Makihara, Y., Muramatsu, D., Echigo, T., Yagi, Y.: Multi-view large population gait dataset and its performance evaluation for cross-view gait recognition. IPSJ Trans. Comput. Vis. Appl. **10** (2018). https://doi.org/10.1186/s41074-018-0039-6
38. Tan, D., Huang, K., Yu, S., Tan, T.: Efficient night gait recognition based on template matching. In: Proceedings—International Conference on Pattern Recognition, vol. 3, pp. 1000–1003 (2006). https://doi.org/10.1109/ICPR.2006.478
39. Thapar, D., Jaswal, G., Nigam, A., Arora, C.: Gait metric learning Siamese network exploiting dual of spatio-temporal 3D-CNN intra and LSTM based inter gait-cycle-segment features. Pattern Recogn. Lett. **125**, 646–653 (2019). https://doi.org/10.1016/j.patrec.2019.07.008
40. Uddin, M., Thanh Trung, N., Makihara, Y., Takemura, N., Li, X., Muramatsu, D., Yagi, Y.: The OU-ISIR large population gait database with real-life carried object and its performance evaluation. IPSJ Trans. Comput. Vis. Appl. **10** (2018). https://doi.org/10.1186/s41074-018-0041-z
41. Wang, L., Tan, T., Ning, H., Hu, W.: Silhouette analysis-based gait recognition for human identification. IEEE Trans. Pattern Anal. Mach. Intell. **25**, 1505–1518 (2004). https://doi.org/10.1109/TPAMI.2003.1251144
42. Wang, R., Shi, Y., Ling, H., Li, Z., Li, P., Liu, B., Zheng, H., Wang, Q.: Gait recognition via gait period set. IEEE Trans. Biometrics Behav. Identity Sci. **5**(2), 183–195 (2023). https://doi.org/10.1109/TBIOM.2023.3244206
43. Wang, Y., Chen, Z., Wu, Q., Rong, X.: Deep mutual learning network for gait recognition. Multimedia Tools Appl. **79**(31–32), 22653–22672 (2020). https://doi.org/10.1007/s11042-020-09003-4
44. Wang, Y., Song, C., Huang, Y., Wang, Z., Wang, L.: Learning view invariant gait features with two-stream GAN. Neurocomputing **339**, 245–254 (2019). https://doi.org/10.1016/j.neucom.2019.02.025
45. Wu, C., Song, Y., Zhang, Y.: Multi-view gait recognition using NMF and 2DLDA. Multimedia Tools Appl. **78**(24), 35789–35811 (2019). https://doi.org/10.1007/s11042-019-08153-4
46. Wu, H., Tian, J., Fu, Y., Li, B., Li, X.: Condition-aware comparison scheme for gait recognition. IEEE Trans. Image Process. **30**, 2734–2744 (2020). https://doi.org/10.1109/TIP.2020.3039888
47. Xianda, G., Zheng, Z., Tian, Y., Beibei, L., Junjie, H., Jiankang, D., Guan, H., Jie, Z., Jiwen, L.: Gait recognition in the wild: a large-scale benchmark and NAS-based baseline (2022)
48. Xu, C., Makihara, Y., Li, X., Yagi, Y., Lu, J.: Gait recognition from a single image using a phase-aware gait cycle reconstruction network. In: Computer Vision—ECCV 2020: 16th European

Conference, Glasgow, UK, August 23–28, 2020, Proceedings, Part XIX, pp. 386–403 (2020). https://doi.org/10.1007/978-3-030-58529-7_23

49. Xu, C., Makihara, Y., Ogi, G., Li, X., Yagi, Y., Lu, J.: The OU-ISIR gait database comprising the large population dataset with age and performance evaluation of age estimation. IPSJ Trans. Comput. Vis. Appl. **9** (2017). https://doi.org/10.1186/s41074-017-0035-2
50. Xu, K., Jiang, X., Sun, T.: Gait recognition based on local graphical skeleton descriptor with pairwise similarity network. IEEE Trans. Multimedia **24**, 3265–3275 (2021). https://doi.org/10.1109/TMM.2021.3095809
51. Xue, W., Ai, H., Sun, T., Song, C., Huang, Y., Wang, L.: Frame-GAN: increasing the frame rate of gait videos with generative adversarial networks. Neurocomputing **380**, 95–104 (2019). https://doi.org/10.1016/j.neucom.2019.11.015
52. Yao, L., Kusakunniran, W., Zhang, P., Wu, Q., Zhang, J.: Improving disentangled representation learning for gait recognition using group supervision. IEEE Trans. Multimedia (2022). https://doi.org/10.1109/TMM.2022.3171961
53. Yeo, S.-S., Rho, S., Kim, H., Safdar, J., Zia, U., Durrani, M.Y.: A triplet-branch convolutional neural network for part-based gait recognition. Comput. Syst. Sci. Eng. **47**(2), 2027–2047 (2023). https://doi.org/10.32604/csse.2023.040327
54. Yu, S., Tan, D., Tan, T.: A framework for evaluating the effect of view angle, clothing and carrying condition on gait recognition. In: 18th International Conference on Pattern Recognition (ICPR'06), vol. 4, pp. 441–444 (2006). https://doi.org/10.1109/ICPR.2006.67
55. Zhang, Y., Huang, Y., Yu, S., Wang, L.: Cross-view gait recognition by discriminative feature learning. IEEE Trans. Image Process. **29**, 1001–1015 (2019). https://doi.org/10.1109/TIP.2019.2926208
56. Zheng, J., Liu, X., Liu, W., He, L., Yan, C., Mei, T.: Gait recognition in the wild with dense 3d representations and a benchmark. In: Proceedings of the IEEE/CVF Conference on Computer Vision and Pattern Recognition (CVPR), pp. 20228–20237 (2022)
57. Zheng, L., Zha, Y., Kong, D., Yang, H., Zhang, Y.: Multi-branch angle aware spatial temporal graph convolutional neural network for model-based gait recognition. IET Cyber-Syst. Robot. **4**(2), 97–106 (2022). https://doi.org/10.1049/csy2.12052
58. Zheng, S., Huang, K., Tan, T., Tao, D.: A cascade fusion scheme for gait and cumulative foot pressure image recognition. Pattern Recogn. **45**(10), 3603–3610 (2012). https://doi.org/10.1016/j.patcog.2012.03.008

Chapter 5
Comparison of the Reviewed Methods

Abstract This chapter provides a comprehensive analysis of methodologies employed in gait identification research, drawing insights from articles gathered in a systematic review spanning 2019–2023. The methodologies are categorized based on their approach: Deep Learning (DL) based and non-Deep Learning (non-DL) based. Traditional non-DL-based methods exhibit limitations in handling large datasets and complex variations. The DL-based methods, leveraging advancements in DL algorithms, demonstrate the capability to extract features automatically, reducing the need for manual feature engineering. Given the predominance of DL-based approaches, a further distinction is made between model-based and model-free approaches. Within DL, model-free approaches are dominant, showcasing a variety of architectures like CNNs, GANs, DAEs, CapsNets, RNNs, and ViTs. These approaches demonstrate strengths in image-based tasks, data generation, unsupervised learning, and sequence modeling. However, model-based approaches have grown recently, particularly with advancements in human pose estimation algorithms. This chapter compares these diverse methodologies, highlighting their strengths, limitations, and contributions to advancing gait identification research.

5.1 Trends in Gait Identification Research

The graphic present in Fig. 5.1 examines the prevalence of DL models in gait recognition literature across the years 2019–2023. It categorizes articles into those that utilize DL-based approaches and those that do not. As DL continues to revolutionize various domains, this graph sheds light on its adoption within gait identification research.

The graph analysis reveals a pronounced emphasis on DL approaches, departing from the earlier methods that leaned more towards traditional machine learning approaches [27]. This trend can be attributed to the remarkable advancements in DL algorithms in recent years. DL models showcase the ability to automatically extract features from various gait representations, whether based on model-based or model-free approaches. One notable advantage is the capacity to perform end-to-end learning, eliminating the need for manual feature engineering. This shift underscores

D. R. M. Bastos and J. M. R. S. Tavares, *Advances in Gait-Based Identification*, Studies in Systems, Decision and Control 593,
https://doi.org/10.1007/978-3-031-89560-9_5

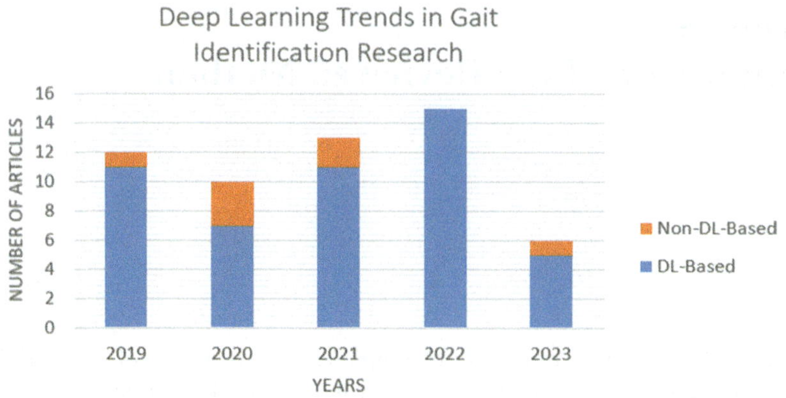

Fig. 5.1 Articles categorization into deep learning-based and non-deep learning-based approaches between 2019 and 2023

the adaptability and efficiency of DL methodologies in gait recognition, fostering a new era of exploration and innovation.

The graphic in Fig. 5.2 illustrates the temporal trends in gait identification research approaches from 2019 to 2023. The analysis categorizes articles into three main groups: those employing model-based approaches, model-free approaches, and a combination of both. By observing the yearly distribution, this graph provides insights into the changing landscape of research methodologies in the field of gait recognition.

The analysis of Fig. 5.2 reveals a distinctive trend in the dominance of model-free approaches in gait recognition models. However, the recent emergence of model-based approaches in the last few years is noteworthy. This surge could be attributed

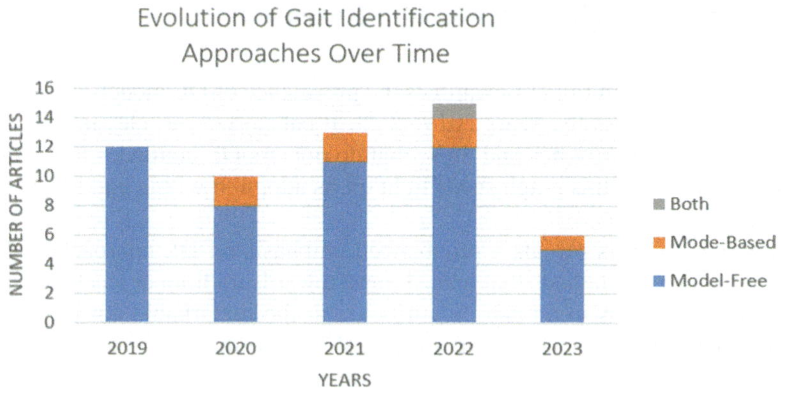

Fig. 5.2 Tracking the trend in gait identification research from 2019 to 2023, this graph highlights the distribution of model-based, model-free, and combined, over the years

to advancements in algorithms related to human pose estimation, a critical component for applying model-based approaches in computational gait recognition. The increased precision and robustness of human pose estimation algorithms may have paved the way for a renewed interest in leveraging model-based methodologies. In 2022, an article was published that encompasses both approaches simultaneously, which could mean the beginning of a new trend for the future [23].

In both graphics, it can be seen that the peak in publications in 2022 stands out, suggesting a significant surge in interest and investigation during that period. A plausible explanation for this spike is the global COVID-19 pandemic, which reached its zenith in 2020 and 2021. The pandemic, emphasizing health and safety measures, prompted intensified research in gait identification as a contactless identification. Moreover, the widespread use of face masks during the pandemic imposed limitations on face recognition technologies, further steering attention towards gait-based identification methods. The dynamic landscape of the research field during these years reflects the adaptability of gait recognition as a valuable tool in evolving circumstances. The articles' publication dates were also considered, providing a chronological representation of the evolving research trends in gait recognition.

The forthcoming sections comprehensively examine gait recognition methodologies, systematically categorized into distinct paradigms. The initial focus scrutinizes non-DL-based models, shedding light on traditional machine learning approaches that have historically played a role in gait recognition. This examination allows the seamless transition to exploring DL-based models, a transformative shift that has reshaped the landscape of gait recognition research. Within the realm of DL, this analysis will differentiate between model-based and model-free approaches. Further granularity is achieved by classifying these articles based on the specific neural network architectures employed. The classification is presented in the graphic of Fig. 5.3. The graphic illustrates the prevalence of articles in different deep architectures classified by their respective approaches: model-based, model-free, or a

Fig. 5.3 Distribution of gait recognition articles across various deep neural network architectures

combination of both. CNNs stand out prominently among them. This approach aims to unravel the diverse spectrum of gait recognition models and illuminate their respective contributions to the field.

5.2 Traditional Gait Recognition Methods

Traditional machine learning approaches in gait identification have evolved with innovative methodologies, each tackling specific challenges and leveraging distinct feature extraction, dimension reduction, and classification techniques. Let's delve into the details of each article:

1. Wu et al. [54]: The process begins with the generation of 2D Enhanced Gait Energy Images (2D-EGEI), a fortified version of GEIs reconstructed with 2D Principal Component Analysis (2D-PCA) to extract primary features while retaining local structured information. This serves as input for the View Transformation Model (VTM), transforming one view angle into another specific view, as it has demonstrated good capability in handling view transformation procedures over the past decades. Non-negative Matrix Factorization (NMF) compensates for the accuracy loss which can learn local information from input features. Additionally, 2D Linear Discriminant Analysis (2D-LDA) is introduced to enhance classification capability by finding a projection matrix that maximizes inter-class scatter and minimizes intra-class scatter. The final classification employs a nearest neighbor classifier. Results indicate a significant improvement in multi-view gait recognition performance compared to traditional methods up to the date of publication of this article;
2. Guo et al. [15]: In this gait recognition algorithm, the structure uses the Gabor filter, commonly employed in image processing, to extract features from GEIs. This filter effectively describes the local structure information of an image, considering spatial location, spatial scale, and direction selectivity. Subsequently, the resulting extracted features undergo dimension reduction using Linear Discriminant Analysis (LDA) to enhance computing efficiency and eliminate redundant information, ultimately improving the linear separability of the data. Classification is performed with an enhanced locally coupled Extreme Learning Machine (ELM). The ELM adopts a neural network structure with a single hidden layer where input weights are randomly assigned, generating corresponding output weights. The output layer utilizes Particle Swarm Optimization (LC-PSO-ELM) in this specific algorithm. The effectiveness of the dimension-reduction method LDA is compared to Principal Component Analysis (PCA), demonstrating superior performance on the CASIA-A dataset. Moreover, the LC-PSO-ELM+LDA algorithm proves effective in cross-view tests conducted on CASIA-B;
3. Bekhouch et al. [4]: This approach introduces a new descriptor, a model-based method, by using the variation in the area of the inner region of the lower limbs between legs, focusing solely on reference frames within a gait cycle. The upper

part of the human body is excluded, and two sets of keyframes correspond to two consecutive half-cycles in the left and right terminal swing. Classification is carried out using the KNN classifier with $k = 1$. Despite the influence of various factors, results on CASIA-B confirm the consistency of features extracted from the lower regions for gait identification. However, this method is effective only for 90° views and does not apply to cross-view problems. Tests using an autoencoder for classification were also conducted, but the KNN classifier exhibited superior results;

4. Lima et al. [29]: The initial gait sequence is used to employ a pose estimation method and estimate the joints' coordinates of a person. The result is input for two simple gait recognition approaches: PoseDist and PoseFrame. In PoseDist, feature descriptors are extracted from the pose joints, and nearest neighbor classification is conducted using two different distances, SDTW and Euclidean. Meanwhile, PoseFrame takes joint coordinates as input, normalizes each joint relative to the neck joint, and feeds these normalized joints into a multilayer perceptron (MLP) optimized with softmax. This MLP learns representations for recognition, classifying each frame independently, and temporal aggregation is applied to predict the identity of a probe sample. The Euclidean distance exhibited better results than the SDTW distance in PoseDist, while PoseFrame showed superior and faster results than PoseDist. Cross-view tests conducted on the CASIA-B dataset revealed that, overall, PoseFrame performs better than deep-learning-based methods in small-sample training. As the training set size increases, PoseFrame may be outperformed. Better results are achieved by focusing on end joints, like wrists and ankles, as they remain visible even when covered by clothing, ensuring more accurate pose estimation;
5. Wen and Wang [53]: Frame-by-frame Gait Energy Images (ff-GEIs) [48] are obtained, undergoing dimension reduction after applying sparse linear subspace technology. The essential step is to project the ff-GEI into a low-dimensional subspace through linear or non-linear subspace methods. The resulting features are more closely distributed and separable. A new algorithm based on Support Vector Machine (SVM) is introduced, adopting Gaussian Radial Basis Function (RBF) kernels to achieve cross-view gait recognition. RBF is employed to address the issue of linear inseparability. The choice of the loss function is a crucial part of the SVM classifier, significantly impacting classification results. Evaluations conducted on the OULP dataset demonstrate excellent performance in cross-view gait recognition with a 10-window temporal extension used in ff-GEI construction. Performance in terms of accuracy was only slightly surpassed by other methods in the literature for the CASIA-B dataset;
6. Iwashita et al. [24]: In this article, the authors aim to enhance robustness to variations in walking speed using an improved Mutual Subspace Method (eMSM). They employ 2D-PCA to accurately calculate the covariance matrix directly from image matrices, reducing accuracy loss compared to 1D-PCA. Discrimination capability is heightened by rotating images at various angles, enabling more robust gait feature extraction. These features are then fused using a boosting method. Two projected matrices are obtained for each combination between the

probe and each class of the gallery. Two similarities are calculated for each pair of projected matrices using the Mutual Subspace Method (MSM) with different parameters. Subsequently, a similarity fusion is calculated for each pair of similarities. The final classification is determined by selecting the highest similarity value. The proposed method outperformed state-of-the-art methods for different speeds. The algorithm has not been tested for other types of variants;

7. Sharma et al. [42]: The authors developed an automatic gait identification system, constructing their own dataset where low-resolution frames were processed to create silhouettes. For each silhouette, the direction of motion is standardized, and the outline is obtained to extract the top and bottom projections. The periodic gait cycle is extracted using function correlation between the current and consecutive frame for both top and bottom projections. The cycle is divided into four, and the projection vectors for each quarter-cycle are extracted and averaged. The correlation between the detected gait and the gait of the ID inserted is calculated using binary classification. Match strength is also calculated. This approach is more suitable for gait authentication involving one-to-one verification.

The analysis of traditional machine learning-based approaches for gait identification reveals common trends and characteristics among the reviewed articles. Notably, these methods often employ variations of GEIs or their enhanced versions, such as 2D Enhanced Gait Energy Images (2D-EGEI) or frame-by-frame Gait Energy Images (ff-GEIs), whose creation process is explained later. Feature extraction techniques, including Principal Component Analysis (PCA), Gabor filters, and linear discriminant analysis (LDA), are frequently integrated into these approaches to enhance discriminative information from gait patterns.

Using distance metrics, such as Euclidean distance, with classifiers like k-Nearest Neighbors (KNN) or Support Vector Machines (SVM) is a prevalent theme across these methods. The emphasis on cross-view gait recognition is evident in several studies, demonstrating the relevance of addressing challenges posed by different viewing angles.

Despite the diversity in methodologies, a common limitation is observed in the applicability of these traditional machine learning-based approaches. They tend to exhibit superior performance in scenarios with smaller datasets, where feature engineering and dimension reduction strategies contribute to effective gait pattern analysis. However, as datasets enlarge, these methods face challenges in handling the complexity and variability present in larger datasets.

5.3 Deep Learning-Based Model-Free Approaches

In recent years, DL-based model-free approaches have emerged as powerful techniques in gait recognition. This category encompasses various neural network architectures that excel at learning discriminative features directly from gait data, eliminating the need for explicit modeling of biomechanical structures. This subsection delves into the diverse landscape of model-free approaches, exploring their unique strengths and applications. CNNs have shown remarkable prowess in image-based tasks, including gait recognition. The focus here is on studies that leverage the hierarchical feature extraction capabilities of CNNs to discern intricate patterns within gait images. GANs, known for their ability to generate realistic data, are explored in the context of gait recognition, investigating how GANs can enhance the quality and diversity of gait data for improved model training. DAEs focus on unsupervised learning and dimensionality reduction, capturing latent representations of gait features for improved feature extraction and reconstruction. CapsNets offer a novel perspective on gait representation, with studies exploring how CapsNets capture hierarchical relationships among different components of gait, potentially improving recognition accuracy. RNNs, designed to handle sequential data, are investigated for their application in modeling temporal dependencies within gait sequences, contributing to capturing the dynamic nature of human gait. 3D-CNNs extend the capabilities of traditional CNNs to volumetric data, handling dynamic and spatial aspects of gait for improved recognition. ViTs, gaining prominence in image analysis tasks, are explored for their application to gait recognition, emphasizing how attention mechanisms contribute to capturing relevant information. Combinations of different models, such as 3D-CNN + LSTM, CNN + LSTM, CNN + GAN, CNN + ViT, and CNN + CapsNet, are investigated within hybrid architectures. The focus is on studies showcasing the synergies and benefits of integrating diverse DL approaches. This comprehensive exploration aims to provide insights into the evolving landscape of model-free approaches, shedding light on the intricate architectures propelling advancements in gait recognition research.

5.3.1 Convolutional Neural Network

This subsection dives into the rich landscape of gait recognition methodologies anchored in Convolutional Neural Networks (CNNs), forming a substantial segment within model-free approaches. CNNs, known for their prowess in capturing intricate patterns and spatial dependencies in image-based tasks, play a pivotal role in the articles discussed here. Each article explores the power of CNNs for feature extraction, unraveling their capabilities in discerning unique gait signatures. Notably, if an article does not explicitly delve into CNNs, it offers insights into alternative convolutional techniques, a deliberate inclusion to encompass the broader spectrum of convolutional approaches. This comprehensive approach ensures coverage of techniques that harness the strengths of convolutional operations for gait feature extraction. As one navigates through this subsection, each article's distinctive methodology is explored, showcasing the pivotal role of CNNs and convolutional techniques in advancing gait identification research. Let's delve into the details of each article:

1. Hawas et al. [20]: This article uses silhouettes to calculate the gait cycle and construct GEIs. From these GEIs, Optical Flow (OF) is computed, highlighting how different body parts move during the gait cycle. A CNN is employed as

a feature extraction based on OF, and a fully connected neural network serves as the classifier. The study demonstrates high recognition rates under various conditions, as clothing and bags were omitted during the OF process, considering them as static parts during walking;
2. Linda et al. [31]: This article introduces a Color-mapped Contour Gait Image (CCGI) to address cross-views. A contour tracing algorithm, the Combination of Receptive Fields (CORF), extracts the first contour in each gait image for key frames using the Difference of Gaussians (DoG) and hysteresis thresholding. In the second phase of CCGI construction, color information encodes spatial and temporal details. This results in more regularized contour images, minimizing discrepancies or unwanted features. This feature representation is fed into a DCNN with softmax as the loss function. Experiments on the CASIA-B database have demonstrated that CCGI can achieve a comparable recognition accuracy of 95.46% in normal walking conditions without cross-view evaluation settings;
3. Wang and Zhang [49]: In this approach, two types of Two-Branch CNNs (TCNN) are used: a middle-fusion TCNN and a last-fusion TCNN. As input for the TCNNs, a new gait feature representation is proposed: Multi-frequency Gait Energy Image (MF-GEI) for one branch and the corresponding GEI for the other in both TCNNs. MF-GEIs are generated by sampling gait energy at different frequencies. Sets of GEIs are created for each frequency and combined, providing a comprehensive representation of gait that incorporates variations at various frequencies. The output of the TCNNs is used as input to the SVM for gait classification. TCNN-based methods show better performance than others, with the last-fusion TCNN demonstrating better performance than the middle-fusion TCNN;
4. George et al. [11]: The authors developed a Siamese Neural Network (SNN) comprising two parallel DCNN, sharing parameters. The network is trained with the Euclidean contrastive loss function. The DCNN takes as input a color-mapped contour gait image (CCGI) computed by averaging different edges detected in the gait silhouette sequences. CCGI is interpolated with color-mapping functions obtained with the edge detection algorithm CORF. The proposed CCGI gait template shows better visual variations than other gait templates. The results also show that the SDCNN can efficiently discriminate gait features, especially in co-factors such as clothing and carrying conditions. Experiments were conducted in gait recognition for fixed-view, cross-view, and multi-view conditions;
5. Wang et al. [50]: Two CNNs are presented in this article that learn together through independent frames of gait, are unaffected by frame alignment, and can work with incomplete gait cycles due to occlusions. Each silhouette serves as input to a convolutional and pooling layer, and the set-level features are obtained. This result is used as input to two neural networks to extract features. At the output of the two CNNs, the Horizontal Pyramid Mapping (HPM) component involves horizontal pooling at different scales and mapping to the discriminative space through fully connected layers. The model is supervised by Triplet Loss, resulting in a balanced loss function between the networks. Performance

evaluation showed state-of-the-art results for the OULP dataset until the date of publication of the article;

6. Fan et al. [9]: The authors propose a novel part-based model named GaitPart. It starts with a sequence of silhouettes as input for the frame-level part feature extractor (FPFE), a special but concise stacked CNN. The resulting feature map is split into four pre-defined horizontal portions. Then, average and global max pooling layers are used, resulting in four sequences of part-level spatial features. Each sequence is input to a micro-motion capture module (MCM) to capture short-range patterns between consecutive frames for a specific part. Frame by frame, a different MCM with different parameters is used to map the frame-level part-informed feature vector into the micro-motion feature vector. The MCM uses a 1D convolutional kernel to re-weight the feature vector at each moment, functioning as a channel-wise attention mechanism. The MCM process concludes with a final temporal pooling. The final gait representation is obtained by concatenating the four MCMs' outputs. Identification is done with the lowest Euclidean distance between the corresponding feature vectors. This approach showed state-of-the-art results on CASIA-B and OUMVLP datasets, serving as a benchmark for subsequent articles;
7. Chao et al. [6]: This article suggests a model-free approach called GaitSet, widely referenced and used as a comparison method in articles related to gait identification. This approach offers flexibility, accommodating diverse inputs without constraints other than silhouette size. It accepts sets of nonconsecutive silhouettes captured from varied viewpoints and walking conditions. The model efficiently learns deep gait representations, requiring one-time computation per sample instead of measuring the similarity between pairs, enhancing recognition speed and efficacy. It starts with a CNN that seeks to extract frame-level features from each gait silhouette. These frame levels are then mapped to set-level features using a permutation-invariant function based on set pooling (SP). The SP uses max() as a statistical function. Another important component is the Multilayer Global Pipeline (MGP), which can be regarded as an additional branch aggregating hierarchical set-level features. The process ends with a Horizontal Pyramid Mapping (HMP) (coming from both branches) that learns the deep set discriminative representation from the set-level feature, and a fully connected neural network is applied to the final classification. In the training process, cross-entropy loss and triplet loss, two widely used loss functions in the identification field, are combined. Experimental findings reveal that the single-model approach achieves notable performance, with an average rank-1 accuracy of 96.1% on the CASIA-B gait dataset and 87.9% on the OUMVLP gait dataset under standard walking conditions. In challenging scenarios such as walking with bags or wearing coats, the model demonstrates robustness, achieving accuracies of 90.8 and 70.3%, respectively, surpassing existing methods on CASIA-B. Notably, the proposed method maintains high accuracy with a limited number of frames, achieving 85.0% on CASIA-B with only seven frames available;
8. Jameel and Dhannoon [25]: The authors built a system tested with a dataset collected by them, containing eight different individuals and eighteen different

videos, corresponding to more than one direction for each person. They begin with the pre-processing of frames to a grayscale. Then, the authors apply a Gaussian function to blur the image, perform image subtraction to remove the background and end with the binary representation of the silhouette. The human body detection process is used to speed up the classification phase. Classification is made with a CNN with a softmax function in the last layer. They achieved an accuracy of 99.5%. This article does not compare the results with other methods on other known datasets, but it is interesting to have an example of how a basic full system of gait identification works;

9. Marin-Jimenez et al. [32]: Existing multimodal gait recognition systems face challenges with body shape changes and dynamic backgrounds that fail when a modality is missing. Addressing this, UGaitNet is introduced as a unified framework resilient to missing modalities, accommodating four different inputs like pixels' gray value, optical flow, depth maps, and silhouettes, regardless of the camera used. It starts with a CNN per input modality in the training process, using a binary value to indicate if the input is available at test time. The information from each branch is merged, creating a multimodal signature. It is followed by a series of fully connected layers with a softmax layer at the end for the final classification. In the TUM-GAID dataset, UGaitNet achieved the new state-of-the-art with optical flow, gray values, and depth maps with 99.2%. Evaluating different combinations in the CASIA-B for the cases of missing modalities in cross-view recognition, it is observed that gray values lead to the best performance, containing extra information compared to silhouettes, except for silhouettes with optical flow, which present the best results when subjects wear coats. However, the robustness of the algorithm when a modality is missing was confirmed;
10. Han et al. [19]: The proposed method focuses on enhancing cross-view recognition performance through metric learning, particularly employing the angular softmax loss. This loss function introduces an angular margin to enhance feature separability, while the triplet loss facilitates extracting more discriminative features. Both loss functions are optimized in different spaces, and including batch normalization mitigates the impact of optimizing two distinct losses. Building upon the GaitSet architecture, acknowledged as the most effective in gait recognition until the moment of writing the article, their method surpasses GaitSet's performance under view-angle, bag-carrying, and coat-wearing variations, as demonstrated in experiments on CASIA-B and TUM-GAID datasets;
11. Wang and Hu [47]: For the proposed method, an ff-GEI is used as input, with each frame going through a multi-channel convolutional block attention network (CBAM) channel. Each channel has three convolutional modules and one CBAM, enhancing the ability of feature expression. Then, all information is fused using a multi-channel convolution feature fusion technology. Gait feature vectors are originated by vertically dividing the global feature into two parts and pooling them. With the calculation of the Euclidean distance between the probe's features and those of the gallery, dissimilarity is measured. Accuracy is assessed based on this distance, considering a 'hit' to the lowest distance;

12. Gao et al. [10]: The authors commence the method with human body segmentation at key positions, addressing the most affected and less unaffected parts in a cloth-changing scenario based on anatomical studies. The proposed network obtains robust hybrid part-based spatiotemporal features from the less unaffected body parts, employing three sub-networks that emphasize distinct aspects of gait. One sub-network focuses on extracting local spatial features, another on generating micro-motion temporal features, and the third on global spatial features. The sub-network for local spatial features deviates from traditional architectures by implementing parallel convolutions, progressively constructing stages in parallel. The third sub-network is added to enhance the discriminative nature of local features. Selective involvement of features from shallow layers is crucial for local details, and a pooling operation consolidates them for the entire sequence. Horizontal Pyramid Mapping (HPM) projects the features into a more discriminative space. The sub-network for global spatial features consists of two branches, each including convolutional and pooling stages, with a final pooling operation and HPM projecting features. The sub-network for micro-motion temporal features utilizes max-pooling operations within successive short-range silhouettes, followed by a final pooling and HPM operation. Concatenating the outputs of the three sub-networks results in an effective hybrid gait feature for final classification. This model, evaluated on CASIA-B and OU-ISIR Treadmill B, demonstrates state-of-the-art results in handling the challenging cloth-changing problem and exhibits good performance with a low number of input frames;
13. Mogan et al. [35]: In this article, the authors propose using a pre-trained VGG-16 model to learn deep gait features from a GEI using transfer learning and fine-tuning techniques. The VGG-16 model, an advancement over AlexNet, features more convolutional layers with smaller kernel sizes. Comprising 13 convolutional layers, 5 max-pooling layers, and 3 fully connected layers, the model organizes these layers into five sets. Ultimately, an MLP is used to determine associations among the feature map, carrying out the final identification of the subject. The experimental results across four different datasets demonstrate promising performances;
14. Sayeed et al. [40]: In this article, a simple ten-layer CNN with GEI as input is employed to evaluate and compare its performance with three activation functions: ReLU, LeakyReLU, and PReLU. The CNN with Leaky ReLU activation function achieved the highest accuracy for the classification rate. Leaky ReLU is an activation function that introduces a small, predefined slope, e.g., 0.01, for negative inputs, preventing them from becoming entirely inactive and addressing the vanishing gradient problem associated with traditional ReLU;
15. Aljazaerly et al. [1]: In this study, a novel loss function called Batch Hard Contrastive Loss (BHCn) is introduced, designed to learn an absolute distance for enhanced general verification. Inspired by triplet losses in match-mining, BHCn preserves the identity-agnostic nature of the contrastive loss by selecting the hardest pair of samples for each pair of identities in a batch. By extending this concept to batch learning, BHCn demonstrates its efficacy on a standard

CNN with GEI inputs and on the widely used methods GaitSet and GaitGL in gait recognition. The experimental results indicate that the suggested BHCn surpasses alternative loss functions, including a triplet loss employing batch mining and the traditional contrastive loss. The performance was evaluated using the Equal Error Rate (EER) metric on the CASIA-B and OU-MVLP datasets;
16. Wang et al. [46]: A novel approach, Iterative Gait Period Set (IGPS), inspired by GaitPart, is proposed. Leveraging the periodic nature of walking, the gait period is divided into distinct phases, using gait phase features to enhance frame-level features. Detection of the gait period relies on identifying gaps in the leg region, generating periodic signals. Silhouettes are segmented into established phases, forming a Period Energy Image (PEI), essentially a GEI with silhouettes from each phase. Frame-level part features are initially extracted using a Frame-level Part Feature Extractor (FPFE), based on 2D convolutional layers and part spatial pooling. The resulting features from different phases are processed through the same IGPS module, generating video-level part features. Horizontal Mapping (HM) transforms these video-level part features, producing the final discriminative part features. Evaluations on the CASIA-B and OUMVLP datasets demonstrate state-of-the-art results, particularly in cross-view conditions with normal walking and considering factors like carrying or clothing;
17. Yeo et al. [60]: Employing a triplet-branch CNN to address covariate factors, each branch processes a specific region of a GEI (head, legs, and body) to classify subjects. Three convolutional layers are used for each GEI part, and the resulting feature maps undergo soft attention layers for enhanced refinement. The attention layers generate attention maps, emphasizing discriminative features crucial for subject identification. The ultimate identification decision is made through a probability-based majority voting criterion. Evaluations on the CASIA-B dataset demonstrated the robustness of the model under clothing and carrying conditions.

In gait recognition, using CNNs has marked a significant stride towards robust and accurate identification. This comprehensive review has delved into 17 noteworthy articles that predominantly leverage CNNs or convolutional techniques for feature extraction within model-free approaches. A common thread among these studies is the reliance on CNNs to discern intricate patterns and discriminative features from gait data. Whether addressing cross-view challenges, clothing variations, or carrying conditions, CNNs have demonstrated their efficacy in capturing the essence of gait dynamics. Attention mechanisms, such as soft attention layers, have refined the feature maps further, emphasizing crucial details for subject identification.

Several studies introduced innovative approaches, such as Color-mapped Contour Gait Images (CCGI) and Iterative Gait Period Set (IGPS), showcasing the continual evolution of methodologies within CNN-based gait recognition. The exploration of diverse activation functions, including Leaky ReLU and PReLU, attests to the ongoing efforts to optimize the learning process and enhance accuracy. Furthermore, the incorporation of multimodal inputs, as demonstrated by UGaitNet, reflects a pragmatic approach to address real-world scenarios where varying modalities might be missing or incomplete in systems with multimodal inputs. The performance gains achieved by iterative methods like Batch Hard Contrastive Loss (BHCn) highlight the importance of refining loss functions for enhanced feature separability. The studies underscored the significance of robustness and speed in gait recognition systems. Methods like GaitSet and GaitPart demonstrated commendable accuracy, even under challenging conditions such as clothing changes and variable frame numbers.

5.3.2 Generative Adversarial Network

A Generative Adversarial Network (GAN) is a powerful DL architecture introduced by Goodfellow et al. [13]. It consists of two neural networks, a generator, and a discriminator, engaged in a game-theoretic framework. The generator aims to generate realistic data, such as images, while the discriminator tries to differentiate between real and generated data. This adversarial process results in the generator continually improving its ability to produce data indistinguishable from real data. GANs have been employed in gait recognition to enhance feature representation and overcome challenges like covariate variations, viewpoint changes, and dataset bias. Within the realm of GAN-based gait recognition, this analysis delves into three notable studies, exploring each of these contributions individually:

1. Wang et al. [51]: In this article, a Two-Stream Generative Adversarial Network (TS-GAN) was proposed to learn view-invariant gait features by creating a GEI of 90° from a GEI from other angle views. This was achieved through a global and partial stream, where the partial stream split the GEI into three parts to learn more local features, while the global part focused on learning more global features. Both generators used the Multi-scale Context-Aware Network (MSCAN) and a pixel-wise loss. Each generator had its own discriminator with the entropy loss. The combination of global and local streams was used to train a total identity discriminator, employing late fusion and softmax, which proved capable of learning better than both separately. The results show that this method performs well on recognitions with large view differences, while it performs worse on datasets with many subjects, indicating generalization issues. They also demonstrated better results with MSCAN and the pixel-wise loss;
2. Gupta and Chattopadhyay [17]: The authors developed a new pose-based gait recognition approach called Dynamic Gait Energy Image (DGEI). This approach is based on a novel method of mapping different frames into pre-determined unique poses. Silhouettes associated with the same key pose are averaged to generate the final DGEI feature. The DGEI is input to a GAN that generates a new DGEI without covariate conditions, making the training and test sequences comparable. PCA is applied for dimensionality reduction, and classification is performed with a LDA classifier. The LDA provides a pose-level prediction for each key pose, and the average of these predictions determines the final identity.

The new pose-based approach proved to be competitive compared to other pose-based approaches. When combined with the GAN, the results were effective in removing covariate conditions such as carrying bags or wearing coats, especially on the CASIA-B and TUM-GAID datasets;

3. Gupta and Chattopadhyay [18]: The authors propose a new gait feature representation called the pose-based Boundary Energy Image (BEI). Following a pose-based strategy, it corresponds to the average silhouette contour information corresponding to several fractional parts of a gait cycle. With at least half of the gait, the silhouettes are split into six different established poses, the silhouette contours are extracted, and the BEI is constructed. The number six for the optimal key poses was found using a k-means clustering algorithm. Considering that a particular frame was associated with a specific pose, the next frame can only be associated with the same key pose as the previous frame or the next key pose. The pose-based BEI undergoes dimensionality reduction with PCA, followed by a final LDA-based classification. Evaluations on CASIA-B and TUMGAID showed better accuracy and classification times than other approaches. This non-deep learning-based approach was also tested with the pose-based BEI features as input to a GAN to remove covariate objects. Evaluations without cross-view conditions, conducted on CASIA-B, showed an increase in performance of just 0,81% with GAN, and the time of classification rose from 1.72 to 3.04 s. Tests on the TUM-GAID dataset also yielded similar results.

Integrating Generative Adversarial Networks (GANs) into model-free gait recognition approaches has proven to be a transformative factor. The introduction of innovative approaches, such as the Two-Stream GAN, Dynamic Gait Energy Image (DGEI) combined with GAN, and the pose-based Boundary Energy Image (BEI) with GAN integration, collectively highlights the versatility of GANs in enhancing feature representation and addressing key issues in gait recognition. These studies collectively showcase the capacity of GANs to generate synthetic data, effectively bridging gaps in dataset bias and enabling robust feature learning. Incorporating GANs in these model-free gait recognition methodologies demonstrates their effectiveness in handling real-world challenges, such as variations in clothing, carrying conditions, and diverse viewing angles.

5.3.3 Deep Autoencoder

Deep Autoencoders (DAEs) represent an inspiring approach to feature learning, leveraging an encoder-decoder architecture to capture intricate patterns within data. The encoder plays a pivotal role in compressing input data into a lower-dimensional "bottleneck" or "code". This compressed representation serves as an abstract and efficient summarization of essential features. On the flip side, the decoder reconstructs the original input from this compressed code. The training process is unsupervised, emphasizing the minimization of reconstruction errors. In gait recognition, DAEs offer a powerful tool for distilling complex gait patterns into a compact and meaningful representation. This adaptive feature extraction proves crucial for robust performance across diverse gait recognition challenges. Notably, a standalone encoder can also function as a DAE, focusing on feature extraction without the reconstruction aspect. Within this framework, six studies exemplify the efficacy of DAEs in addressing various challenges in gait recognition:

1. Babaee et al. [3]: The authors developed an algorithm to address occlusion issues, particularly in scenarios where only a few frames of a full gait cycle are available. The process begins with obtaining the incomplete Gait Energy Image (IC-GEI) using the available frames. A gradual transformation to a complete GEI (RC-GEI) is achieved through an end-to-end network called ITCNet, essentially an encoder-decoder. The encoder section integrates the intermediate outputs of nine Fully Convolutional Neural Networks (FCNs), with each FCN operating on a specific part of the gait cycle, covering 1/10 of the total cycle length. This progressive transformation from IC-GEI to RC-GEI is followed by PCA, and the final classification is carried out using the nearest neighbor classifier. The evaluation using the CASIA-B and OULP datasets demonstrates that ITCNet significantly improves the recognition rate, especially when only 0.1 part of the gait cycle is available. However, it has not been tested under cross-view conditions;
2. Li et al. [28]: The authors propose the unified joint intensity transformer network (JITN) as the first to consider joint intensity and spatial metric learning. The network is a unified CNN-based architecture consisting of three parts. The first part, Joint Intensity Metric Estimation Net (JIMEN), is essentially an encoder-decoder network that estimates the joint intensity metric. The second part is the joint intensity transformer, responsible for applying a joint intensity transformation of a pair of GEIs. The final part, the Discrimination Network (DN), is a generic CNN that receives the spatial dissimilarity image to learn the spatial metric and perform identification using the triplet loss function. This algorithm exhibits the lowest equal error rate (EER) compared to other existing approaches for all four datasets used. Additionally, it has proven to be resilient to noise and demonstrated state-of-the-art results, improving robustness against variations such as clothing and carrying status;
3. Gupta and Chattopadhyay [16]: The authors suggest improving pose-based gait recognition approaches, specifically addressing frame rate and walking speed variations. The framework involves a gallery of key pose sets corresponding to different walking speeds, enhancing robustness. The key frame is initially established, and each subsequent frame is associated with a key pose from the pre-determined key pose sets. A key pose-based Active Energy Image (AEI) is computed for each silhouette, involving pixel-wise absolute differences between adjacent silhouettes. A Generalized Active Energy Image (GAEI) is obtained by averaging the difference image sequence associated with a specific key pose. This process is repeated for each key pose in the key pose set. The GAEIs are fed through an autoencoder neural network to obtain a condensed representation in a lower-dimensional space. The test features are then projected into this shared space,

and an LDA classifier is employed for the final classification. This pose-based method has proven efficiency in recognizing people with carrying and clothing conditions in the CASIA-B and TUMGAID datasets.

4. Wu et al. [55]: The proposed solution comprehensively addresses cross-condition challenges in gait recognition. It introduces the Instructor model for adaptive silhouette comparison, leveraging local part re-weighting. The Dresser enhances model robustness through blurring and thresholding transformations, contributing to enriched data variation. Temporal modeling benefits from the Dilated Temporal Pyramid Convolution (DTPC), featuring an encoder structure for efficient multi-time span feature extraction. These components synergize in a two-stage training process. In Stage A, Dresser and DTPC process silhouette sequences for local features used in triplet loss for gait representations. Stage B involves concatenating sequences for dynamic weight generation by the Instructor, enhancing cross-condition adaptability in the comparison scheme. During inference, the dynamic comparison scheme calculates sequence similarity using features generated by the DTPC. The final classification is determined by pairing samples with the highest similarity, showcasing superior performance, especially in cross-condition scenarios. The Instructor's dynamic weight analysis reveals a focus shift from the shoulder area to the head or feet with increasing view differences;
5. Hou et al. [22]: A Gait Quality Aware Network (GQAN) was proposed, consisting primarily of two blocks: a Frame Quality Block (FQBlock) and a Part Quality Block (PQBlock), aimed at assessing the quality of each silhouette and each part, respectively. The network takes a sequence of silhouettes as input, where each silhouette undergoes an encoder to extract features individually before passing through the FQBlock. The FQBlock horizontally slices the features of each silhouette, applies global max and average pooling to compress channel information, and utilizes two fully connected layers for excitation, adaptively recalibrating the features for each silhouette. A set-level representation is then obtained through set pooling, aggregating features in an unordered set using max pooling. Subsequently, a PQBlock is employed to learn adaptive weights for each part to match gait sequences. This involves a fully connected layer predicting scores to assess the relative importance of each part. The network is trained end-to-end with a part quality loss. The final classification is determined by computing the Euclidean distance between the probe and gallery, with parts adaptively weighted. The GQAN was evaluated on CASIA-B and OUMVLP datasets, demonstrating state-of-the-art results in cross-view conditions and various scenarios involving different clothing and carried objects. The article also highlights the challenge of interpretability in gait recognition, claiming advancements toward enhancing interpretability for silhouette-based images;
6. Yao et al. [59]: The proposed network follows a basic encoder-decoder framework, taking input from four sequences of frames, each corresponding to a different gait acquisition. One sequence serves as the anchor and the remaining four act as references. The method is order-independent, with the anchor sequence sharing at least one common attribute with each reference sequence. The encoder processes each frame, breaking down the encoded feature representation into

latent features for pose, gait, appearance, and view. To ensure the adaptability and independence of features, a disentanglement swap module is integrated into the encoder-decoder process, using swap operations based on shared latent features between the anchor and each reference. A disentanglement aggregation module enhances the resilience of disentangled gait and appearance features. Finally, initially derived from individual frames, these features are concatenated for the identification process. The algorithm's performance was evaluated using both silhouettes and RGB frames. The results showed better results for RGB frames in normal walking and carrying conditions, but as expected, silhouettes demonstrated much better accuracy in cloth-changing conditions. These experiments verified that this proposed method can achieve prominent results.

The six studies highlighted within this subsection showcase the versatility and effectiveness of DAEs in addressing various challenges. These challenges range from occlusion issues and frame rate variations to clothing and carrying conditions. The application of DAEs has demonstrated the capability to handle covariant features, allowing for extracting a representation without these features with the encoder-decoder process. An encoder-decoder process can also convert viewing angles, providing an avenue for obtaining representations from different perspectives. The compressed representations can also be directly used in classification tasks. This compressed representation is a robust and efficient summarization of essential features, offering adaptability and resilience across diverse gait recognition challenges.

The ability to distill complex gait patterns into compact and meaningful representations while also handling variations in real-world conditions positions DAEs as a valuable tool for advancing state-of-the-art gait recognition technology.

5.3.4 Capsule Network

Capsule Networks, or CapsNets, represent a distinctive paradigm in DL architectures that diverges from traditional Convolutional Neural Networks (CNNs). Introduced to overcome some of the limitations of CNNs in handling hierarchical relationships and spatial hierarchies, CapsNets were proposed by Sabour et al. [39]. The fundamental building block of a CapsNet is the capsule, a group of neurons designed to encode specific instantiation parameters of an entity, such as the pose or orientation. Unlike neurons in a traditional neural network layer, capsules consider not only activation but also the existence and properties of entities in higher layers. Capsules facilitate dynamic routing, allowing the network to learn the hierarchical spatial relationships between different parts of an object. The routing process helps CapsNets capture intricate patterns, especially in scenarios where spatial hierarchies are crucial. This unique architecture holds promise for various applications, including image recognition and gait identification in the current context. The subsequent study navigates the application of CapsNets in gait identification, particularly in challenging scenarios such as cross-view, clothing-changing, and cross-walking conditions, using two distinct capsule network-based architectures.

One focuses on matching local features at the bottom layer (LBC), while the other emphasizes matching mid-level features at the middle layer (MMC). Tests were conducted on the CASIA-B and OU-ISIR Treadmill dataset B. MMC's overall performance proved superior to LBC for the normal, bag, and clothing conditions in the CASIA-B dataset. Regarding the depth of the network as a key factor, the authors concluded that it had no obvious influence on the normal condition, while a four-capsule layers network exhibited better performance in clothing and bag conditions. The impact of the input image was also studied, comparing GEI and Chrono-gait image (CGI). The study suggests that GEI appears to be more suitable for the proposed network. In conclusion, the authors demonstrated that input images with a resolution of 126×126 pixels achieve slightly higher accuracy than input images of 224×224 pixels. This CapsNet exhibits robustness to varying view angles, walking conditions, and clothing variations [57].

Compared to conventional CNNs, CapsNets have emerged as a promising alternative, showcasing distinct advantages in addressing challenges associated with viewpoint changes and data variability. CapsNets offer the potential to capture more nuanced spatial relationships between different body parts, a critical aspect of gait recognition. Despite these merits, it is essential to acknowledge certain limitations. One significant drawback lies in the high computational cost of CapsNets, rendering them less suitable for real-time applications [41]. The computational expense of CapsNets, coupled with a potential demand for more extensive training data compared to CNNs, poses challenges for practical deployment. Additionally, the interpretability of CapsNets introduces another layer of complexity. Unlike CNNs, CapsNets represent their outputs as vectors of probabilities rather than feature maps, making interpreting results more intricate. Navigating these considerations is crucial for understanding the trade-offs and effectively harnessing the strengths of CapsNets in the dynamic landscape of gait recognition research and applications.

5.3.5 Recurrent Neural Network

Recurrent Neural Networks (RNNs) are a potent class of neural network architectures designed for handling sequential data, featuring loops within their structure to maintain information from previous steps. While effective for capturing temporal dependencies, traditional RNNs face challenges with long-term dependencies. Addressing this, Long Short-Term Memory (LSTM) networks offer a solution by incorporating specialized memory cells and gating mechanisms. LSTMs, equipped with input, forget, and output gates and sigmoid and hyperbolic tangent activations, excel at learning long-term dependencies. In gait identification where understanding temporal dynamics is crucial, using RNNs and LSTMs becomes paramount [61].

The subsequent discussion thoroughly explores gait identification methodologies, focusing on studies in the systematic review that harness the capabilities of RNNs and LSTMs. These advanced DL architectures play a pivotal role in extracting intricate temporal patterns from gait sequences, significantly contributing to the evolving landscape of gait recognition research:

1. Wang and Yan [48]: This study employs a novel gait feature representation, ff-GEIs, as inputs to the Conv-LSTM, which enhances the traditional LSTM model with added convolutional layers. This choice is motivated by the time series nature of human gait, where each step is influenced by its preceding state. A final softmax layer converts LSTM outputs into probabilities associated with each class. The ff-GEIs result from combining normalized templates of silhouettes at specific time intervals. Subsequently, these ff-GEIs are sequentially fed into the LSTM. Each ff-GEI represents a specific window of frames in the gait sequence, integrating temporal and spatial features. This is achieved by combining adjacent frames at defined intervals, creating multiple ff-GEIs capturing unique perspectives of the gait over time. The algorithm exhibits advantages over GEI, expanding the available and trainable gait data while relaxing the constraints of gait cycle segmentation. The model was trained on CASIA-B and tested on OULP to assess its generalization capability;
2. Moghaddam and Etemad [37]: The proposed attentive recurrent model initiates by learning convolutional maps from individual frame-level gait silhouettes. To capture spatio-temporal information, a temporal pooling layer aggregates the frame-level representations, yielding the Gait Convolutional Energy Map (GCEM). GCEM, invariant to different viewing perspectives, is then divided into bins, serving as input to a fully connected layer to reduce dimensionality, resulting in a sequence of partial features. A network of bi-directional gated recurrent units captures relationships between spatio-temporal components crucial for gait recognition. An attention layer selectively emphasizes essential recurrent representations, and a softmax layer produces final features for classification using a cosine proximity loss function. The model demonstrates robustness in cross-view settings, different clothing and carrying conditions, and against various synthesized occlusions;
3. Wen and Wang [52]: Cross-view gait recognition based on residual long short-term memory, specifically CVGR-RLSTM, is achieved using ff-GEIs. The CVGR-RLSTM architecture starts with a residual block, based on three convolutional layers, to further refine spatial features, followed by two LSTM blocks to optimize temporal gait features. Shortcut connections are added, transforming the network architecture into a residual version. Shortcut connections, introduced in Residual Networks (ResNets) [21], facilitate gradient flow during DL training. Evaluations on CASIA-B and OULP reveal that the proposed Res-LSTM method outperforms other methods in terms of correct recognition rate;
4. Ghosh [12]: The proposed method begins with applying convolutional layers to GEIs obtained frame by frame. The feature matrix generated from the last convolutional layer is used to produce regions of interest (ROIs) through a single region

proposal network (RPN), an integral part of the faster region convolutional neural network (R-CNN), a class of CNN architectures used for object detection tasks. Faster R-CNN detects and extracts only the pedestrian from frames without carrying objects. A feature vector is generated using deep convolutional layers, and a bidirectional long-short-term memory (BLSTM) studies this feature vector to recognize individual walking patterns, concluding with classification. The BLSTM employs two LSTM layers, one processing the sequence progressively and the other regressively, allowing it to capture contextual information bidirectionally compared to LSTM, which captures temporal dependencies in a single direction. The pedestrian detection algorithm exhibits performance above 98.74% for four datasets. Results also show that BLSTM outperforms LSTM. Using Faster R-CNN to detect only the pedestrian when carrying an object demonstrates the ability to generate similar vectors from the walking patterns of the same pedestrian without objects. This generative approach yields state-of-the-art results.

Using RNNs and LSTM networks in gait identification proves to be a transformative approach. Incorporating these advanced DL architectures allows for extracting intricate temporal patterns from gait sequences, addressing challenges associated with time series-based data. The studies presented in this section showcase diverse applications, ranging from refining spatial and temporal features to achieving robustness in cross-view settings, handling variations in clothing and carrying conditions, and even surpassing the limitations of traditional gait representations like the GEI. The introduction of novel features, such as ff-GEIs and Gait Convolutional Energy Maps (GCEMs), demonstrates the adaptability of RNNs and LSTMs in capturing essential gait characteristics. Additionally, the employment of residual architectures and attention mechanisms further enhances the performance of these models, leading to superior recognition rates. Overall, incorporating RNNs and LSTMs in gait identification advances the state-of-the-art and opens avenues for exploring deeper temporal dependencies and nuanced spatiotemporal representations in this domain.

5.3.6 Three-Dimensional Convolutional Neural Network

Three-dimensional Convolutional Neural Networks (3D-CNNs) represent a notable advancement in the domain of gait identification, extending beyond the capabilities of traditional 2D-CNNs by incorporating an additional dimension. While 2D-CNNs excel at processing spatial information in images, 3D-CNNs go further by considering the temporal evolution of features over consecutive frames. This added temporal dimension allows 3D-CNNs to capture the dynamic aspects of gait sequences, making them well-suited for tasks involving appearance and motion information. In gait recognition, this temporal awareness becomes crucial as it enables the model to discern subtle variations in gait patterns across different frames. Now, delving into the exploration of 3D-CNNs for gait identification, three studies are highlighted, each offering unique insights into the use of this architecture:

1. Castro et al. [5]: This article conducts a comparative study on various CNN architectures using gray pixels (providing appearance information), optical flow channels, capturing gait-related motion, and depth maps, offering soft geometric information. The authors design 2D-CNN and 3D-CNN architectures with different depths and evaluate various early or late fusion methods to combine information from the three modalities. The evaluation on CASIA-B and TUM-GAID suggests that raw pixel values serve as a competitive input modality, outperforming traditional silhouette-based features like GEI. The fusion of information from gray level pixels, optical flow, and depth maps yields state-of-the-art results with significantly smaller image resolution than the referenced reports, except for cases involving clothing variance. The study emphasizes the crucial influence of CNN selection and design on the final results;
2. Gul et al. [14]: This study employs a condensed gait feature representation in the form of a colored GEI as input to a 3D-CNN to capture spatio-temporal gait features. The inclusion of image color channels adds an extra dimension to the three-dimensional representation of the GEI. Fully connected layers with a softmax final layer are used for the final classification. Through exploration of optimization using Bayesian algorithms to tune hyperparameters, the research achieves state-of-the-art results on CASIA-B;
3. Lin et al. [30]: This article introduces a novel global-local-based gait recognition network named GaitGL. Comprising three stages of dual-branch structures, the network effectively leverages global and local information from gait feature maps. The first branch, a global feature extractor, captures contextual information about the relationships among various body parts. Meanwhile, the second branch, the mask-based local feature extractor, delves into the detailed posture changes of local regions. The mask-based partition strategy randomly generates complementary mask pairs to occlude gait sequences for training the local feature extractor, incorporating part-level, strip-level, and pixel-level partitions. Three 3D convolution operations are performed in each stage, extracting global features from the entire sequence of silhouettes in one and local features from the randomly generated complementary mask pairs in the others. The outputs from the three 3D convolutions are fused, creating a new 3D block representing the entire gait sequence, which then undergoes the same process in the next stage. The first stage concludes with local temporal aggregation, followed by spatial pooling at the end of the second stage. Finally, in the third stage, the network performs the final classification by calculating the Euclidean distance between the feature of each sample in the probe set and all features of samples in the gallery set. GaitGL is the first network to exploit global and local information in gait sequences fully, achieving state-of-the-art results across four well-known datasets. This article is widely used as a reference and comparison of results by other articles.

The studies presented within this subsection collectively demonstrate the robustness of 3D-CNNs in addressing challenges such as variations in view angles, clothing changes, and walking conditions. The fusion of appearance information, gait-related motion, and soft geometric details through 3D-CNN architectures has yielded state-of-the-art results, surpassing traditional silhouette-based features. Furthermore, the adaptability of 3D-CNNs to input modalities, such as colored Gait Energy Images (GEIs), showcases their versatility and capacity to handle diverse representations of gait features. In conclusion, integrating 3D-CNNs into gait identification models marks a significant stride forward, paving the way for more sophisticated and accurate systems in the evolving landscape of gait recognition research.

5.3.7 Vision Transformer

Vision Transformers (ViTs) constitute a recent class of neural network architectures designed for image processing and visual feature extraction [8]. In contrast to traditional convolutional architectures that use convolutional layers to extract local patterns, ViTs adopt a global transformational approach. They apply attention mechanisms across the entire image, allowing the network to learn spatial relationships between different parts of the image without relying on convolutional operations. The basic architecture of a Vision Transformer involves dividing the image into small patches, which are then flattened and treated as linear sequences. Each patch is transformed through linear layers and attention mechanisms. This approach enables ViTs to capture long-range patterns and global relationships in an image, making them effective for computer vision tasks, including pattern recognition in complex visual data.

By integrating Vision Transformers into architectures dedicated to gait identification, these models aim to exploit the capacity of these networks to learn robust representations of gait patterns, contributing to the advancement of the field of gait recognition. The following two articles incorporate Vision Transformers into their approaches to gait identification:

1. The first proposed approach [38] stands out as one of the pioneering efforts to leverage the Vision Transformer (ViT) architecture for gait recognition. Initially, the authors use GEI as input for a self-supervised learning technique known as Data Independent Noise (DINO). DINO involves two networks, a teacher and a student, sharing architecture but with distinct parameters. A multicrop strategy is employed to generate both global and local views from input images. The similarity between network outputs is measured using cross-entropy loss, with student parameters optimized to minimize this loss, while teacher parameters are maintained as an exponential moving average of the student's parameters. DINO enables the model to learn valuable gait features without relying on specific labels. Based on the ViT architecture, the pre-trained feature extractor is used alongside a fully connected neural network classifier for gait recognition. This strategy enables the model to automatically acquire meaningful representations of gait patterns, contributing to improved performance in gait recognition tasks.

While this network excels in normal conditions, it exhibits lower accuracy when confronted with changing clothes;

2. The second proposed approach [36] addresses the absence of attention mechanisms in conventional gait recognition architectures, such as CNNs, by introducing the Vision Transformer (ViT), which is termed Gait-ViT. Operating on GEI involves dividing the image into patches, sequentially converting them through flattening and patch embedding, and jointly applying position and patch embeddings to preserve spatial information. The sequence of vectors transforms within a Transformer encoder, leveraging multi-head attention mechanisms, receptive fields, layer normalization, global operations, and residual connections. This comprehensive methodology significantly enhances the model's ability to encode essential gait features, elevating overall recognition performance. The final classification employs a multi-layer perceptron (MLP) with the cross-entropy loss function. The authors report impressive accuracy figures, achieving 99.93% on CASIA-B, 100% on OU-ISIR TreadMill D, and 99.51% on OULP. However, details regarding test conditions, evaluation metrics, and data partitioning are omitted, hindering a direct objective comparison with other articles in the literature.

The exploration of ViT methods in gait identification signifies a notable advancement in deep learning-based model-free approaches. These approaches demonstrate the efficacy of ViT architectures in capturing and understanding intricate gait patterns. Leveraging self-supervised learning techniques and overcoming the limitations of traditional architectures like CNNs, ViT-based models, such as Gait-ViT, showcase the ability to automatically learn valuable representations of gait features. The multicrop strategies, attention mechanisms, and patch-based processing incorporated in ViT architectures enhance recognition performance.

5.3.8 Hybrid Deep Architecture

Hybrid deep architectures in gait identification represent innovative combinations of distinct neural network components, each contributing specialized capabilities to enhance the overall model's performance. The following section explores five notable hybrid architectures, each uniquely tailored for gait recognition. These combinations include the integration of 3D-CNN with LSTM, CNN with LSTM, CNN with GAN, CNN with ViT, and CNN with CapsNets. These architectures harness the strengths of both constituent models, leveraging spatial and temporal features, generative capabilities, and attention mechanisms to achieve more robust and comprehensive representations for gait pattern analysis. The subsequent discussions unravel the details of each hybrid architecture, shedding light on the distinctive contributions and advancements presented by the selected articles within the systematic review.

Each architectural fusion is explored in the context of the included studies, providing insights into their individual approaches and innovations within the domain of model-free gait identification:

- 3D-CNN + LSTM: The combination between 3D-CNN and LSTM is used on LSTM-VGRNet2 [44], a view-invariant network. The process begins with creating a silhouette stereo map (SSM) by fusing adjacent binary silhouettes and calculating pixel-level differences between two images. SSM captures the magnitude and direction of variation between adjacent silhouettes, which are then split into five gait cycle segments (GCS). For each GCS, spatiotemporal features are extracted using a pre-trained 3D-CNN, the VGRNet, capable of capturing intra GCS features. These results are concatenated temporally and fed into an LSTM cell, learning long-term dependencies between each GCS. The average output corresponds to the class score. LSTM-VGRNet2 is trained end-to-end using a triplet loss function. To achieve view invariance, an accurate viewing angle of the subject is determined by VGRNet. However, the algorithm's effectiveness may be limited if there is no suitable match in the dataset for the specific viewing angle. A comparison with silhouette usage indicates a slight performance increase when using SSM;
- CNN + LSTM: The article Zhang et al. [62] addresses an architecture that combines CNNs with LSTMs but first begins by introducing a novel gait-related loss function called angle center loss (ACL) to learn discriminative gait features. ACL is robust to different local parts and temporal window sizes, learning multiple sub-centres for each angle of the same identity. In contrast to center loss, which learns only one center for each identity, ACL achieves better intra-subject compactness by penalizing only the largest distance between the anchor feature and the corresponding cross-view sub-centers. The algorithm starts by dividing silhouettes into four horizontal parts. Each part undergoes individual CNN processing through a simplified spatial transformer network to extract discriminative spatial features and localize suitable horizontal parts of the human body. Local features with the learned partition outperform global features, emphasizing the importance of the head and upper legs in cross-view recognition. An LSTM unit is used as a temporal attention model for temporal feature learning, focusing on more discriminative frames. The temporal attention model outperforms GEIs. The LSTM output feeds into separate loss functions during training. In testing, weighted features for each part are concatenated, and cosine similarities determine identification. Evaluation across three datasets demonstrates that ACL outperforms other state-of-the-art loss functions;
- CNN + GAN: Between the articles included in this literature review, two articles combine CNN with GAN architecture. The first one is Xue et al. [58], which introduces an innovative method, a new type of GAN, called Frame-GAN, to increase the frame rate of gait videos. Some generated frames exhibit even better quality than real next frames, contributing to noise reduction significantly. Leveraging recent advances in metric learning, a new effective loss function named Margin Ratio Loss (MRL) is introduced, enhancing the recognition model. The classifier

follows the structure of GEINet, an eight-layered CNN with fully connected networks in the last two layers for final classification. Evaluations on the CASIA-B and OULP datasets demonstrate improved frame rate GEI and the effectiveness of GEINet [43] classification with MRL optimization. The second article, Alsaggaf et al. [2], proposes a deep learning approach to addressing the challenge of recognizing individuals under diverse walking conditions. The model initially trains on individuals with normal walking patterns, and to adapt to dynamic conditions, a cycle-consistent generative adversarial network (CCGAN) is used. This CCGAN transforms GEIs affected by different factors into a standard GEI, ensuring the model's adaptability to varying walking scenarios. The approach combines CNNs for the final classification and CCGANs, achieving consistent person recognition in real-world scenarios with changing walking conditions;

- CNN + ViT: The only article combining CNNs and ViT employs a CNN to extract features and subsequently apply a joint attention mechanism (CJAM) to discern the most discriminative areas, achieving person identification [26]. The attention mechanism uses a transformer architecture to use a gait feature representation based on a compressed version of silhouette contours obtained through specific edge detection as input. A specific modification to the transformer for CJAM involves removing its standard decoder and adding a randomly initialized decoder query vector. The initial gait sequence is divided into 5-frame subsequences, processed by a CNN into feature vectors suitable for attention mechanisms. Overlapping subsequences are analyzed, and mean class probabilities are calculated for each. CJAM assigns the class with the highest probability to the entire sequence for final classification. Experiments using the CASIA-A and CASIA-B datasets demonstrate that the new attention model leads to lower error rates, improving upon 3D-CNNs, Conv-LSTM, and simple CNNs. A second article proposes a novel approach that combines CNNs and the Vision Transformer for gait recognition [34]. Using GEI as input, three distinct models, namely DenseNet-201, VGG-16, and a Vision Transformer, are employed to capture distinctive gait features related to a subject. The models undergo pre-training and fine-tuning. Each model generates a prediction score, aggregating the scores by summing and averaging to determine the final class label. Implementation incorporates strategies like early stopping, skip connections, and layer normalization to prevent overfitting and reduce computational costs. This fusion of three models enhances the overall generalization capacity for gait recognition;
- CNN + CapsNet: Merlin Linda et al. [33] presents an approach that commences with converting raw input data into binary silhouettes. The CORF edge detection algorithm and gait period detection are then applied to construct a Color-mapped Contour Gait Image (CCGI). The proposed model, CNN-CapsNet, learns automatic feature representations based on a CNN and utilizes capsule vectors as neurons to encode all spatial information. CapsNet maintains spatial gait features through a routing-by-agreement mechanism and addresses translational invariances by introducing orientation equivariance. This approach overcomes limitations typically associated with CNNs. The final classification involves a fully connected neural network using Adam optimizer and margin loss as the loss function

to train the model. Evaluations using the CASIA-B dataset demonstrate robustness against cross-viewing and variations in transportation and clothing conditions.

In conclusion, exploring hybrid deep architectures in gait identification reveals diverse models designed to address specific challenges inherent in gait pattern analysis. Integrating 3D-CNN with LSTM strategically divides the learning process, enabling the model to discern short-term dependencies with 3D-CNN and long-term dependencies with LSTM within the gait sequence. Similarly, combining CNN with LSTM emphasizes spatial and temporal feature learning, allowing the model to capture spatial intricacies and temporal dependencies in gait sequences.

In the realm of CNN + GAN hybrids, the GAN plays a dual role by enhancing frame rate by adding high-quality frames and removing confounding factors from GEIs. The CNN complements this process by extracting discriminative features and facilitating the final classification. Combining ViT with other architectures, such as CNN, conducts primary feature extraction using CNN and leverages ViT's attention mechanisms to focus on the most discriminative parts of the gait patterns.

Finally, the fusion of CNN with CapsNet demonstrates a promising avenue to overcome limitations associated with traditional CNNs. CapsNet introduces a novel approach to feature representation, offering potential advancements in capturing complex relationships within gait patterns. In essence, these hybrid architectures showcase the synergy of diverse neural network components, contributing to more robust and nuanced gait identification methodologies.

5.4 Deep Learning-Based Model-Based Approaches

The domain of gait recognition has undergone a profound transformation with the integration of deep learning methodologies, particularly within the realm of model-based approaches. This chapter delves into distinct architectural paradigms that have played pivotal roles in advancing gait identification computational systems. The following subsections explore the use of CNN, Graph Convolutional Networks (GCN), Recurrent Neural Networks (RNN), and Vision Transformers (ViT) in modeling gait patterns. Each subsection discusses how these architectures enhance gait identification systems by leveraging skeletal information. The reader is guided through exploring convolutional layers in CNN, the graph-based structures of GCN, the temporal dependencies harnessed by RNN, and the innovative integration of transformers in ViT. This structured examination offers a concise yet comprehensive understanding of each architecture's role in shaping the landscape of gait identification within a model-based framework.

5.4.1 Convolutional Neural Network

Within the domain of gait recognition, model-based approaches leveraging Convolutional Neural Networks have demonstrated robustness against variations induced by dress and speed changes. This subsection explores a notable model employing a Local Graphical Skeleton Descriptor (LGSD) [56].

The LGSD, designed to adapt to varying positions and frame counts, extracts four distinct local patterns: position, angle, swing, and trajectory. The position local pattern encapsulates three relative positions of neighboring joints, elucidating the scale and shape of the body. The angle local pattern computes length and angle relationships between spatial and temporal neighboring joints, capturing both spatial and temporal local angle patterns. The swing local pattern characterizes the intensity and velocity of limb swing by analyzing the change in position between a joint and its parent joint. Lastly, the trajectory local pattern delineates the motion of each joint within a short time interval.

The LGSD's normalization of all distances to resist scale changes in the camera is noteworthy, ensuring adaptability across diverse scenarios. The subsequent integration of the Pairwise Similarity Network (PSN), embedded in a two-stream CNN structure, further enhances the gait recognition process. The PSN incorporates a siamese stream for similarity comparison and a concatenated stream for true/false classification. The network processes histograms representing each gait, resulting from concatenating four LGSD local pattern histograms. These histograms serve as inputs for the PSN, contributing to its ability to discern whether two gaits originate from the same person.

While the algorithm excels in maximizing the similarity of true pairs and minimizing the similarity of false pairs on datasets such as CASIA-A, CASIA-B, and CASIA-C, it exhibits limitations in addressing cross-view recognition challenges. This observation suggests that appearance-based approaches offer greater robustness to view changes than skeleton-based methodologies.

This model-based approach, featuring a CNN architecture, underscores the intricate interplay between local graphical skeleton descriptors and DL frameworks, contributing to the evolving landscape of gait recognition systems.

5.4.2 Graph Convolutional Network

Within gait recognition, Graph Convolutional Networks (GCNs) have emerged as a powerful tool for capturing spatial and temporal dependencies in skeleton sequences. This subsection explores notable contributions within this paradigm, shedding light on using GCNs in two model-based gait recognition approaches.

At its core, a GCN is designed to process data structured as graphs. In gait recognition, a skeleton sequence can be conceptualized as a graph, where joints represent nodes, and the connections between joints form the edges. A GCN operates by iteratively aggregating information from neighboring nodes, enabling the model to discern complex relationships and temporal dependencies within the skeleton sequence. This unique capability makes GCN particularly well-suited for tasks involving spatial and temporal dynamics, aligning seamlessly with the intricacies of gait analysis [64].

The Gait-D model, proposed by Iwashita et al. [24], introduces a skeleton-based recognition approach that leverages GCN. The methodology begins with applying the OpenPose pose estimation algorithm to extract a skeleton sequence consisting of 18 joints in each frame of a raw gait sequence. Spatial and temporal features are extracted by establishing relationships between body joints through a GCN conbined with a Temporal Convolutional Network (TCN). The Spatial-Temporal Graph Convolutional Layers (ST-GCN) consist of three layers, each with a GCN to extract spatial features and a TCN to capture temporal dependencies. The feature map is refined further using Canonical Polyadic Decomposition (CPD), a low-rank decomposition algorithm. The final classification is achieved through a softmax classifier after global pooling. The Gait-D model demonstrates robust gait recognition, showcasing resilience under varied conditions such as bags or cloth coverings, and maintains stability across different views and walking conditions.

In the pursuit of developing a cross-view invariant gait identification algorithm, Zheng et al. [63] present the MAST-GCN model. The model employs the HRNet method to extract the skeleton structure from raw input data, chosen for its proficiency in capturing joint positions in ambiguous and low-resolution videos, as observed in the CASIA-B dataset. MAST-GCN adopts a Spatial Temporal Graph Neural Network (ST-GCN) as its foundation, employing a multi-branch structure with four ST-GCN blocks. These blocks divide skeleton frames into four subgraphs, each processed by a dynamic joint branch and a static bone branch. Each ST-GCN block contains a Spatial Graph Convolutional Network (SGCN) and a TCN to extract features in the spatial and temporal domains. The joint and bone branches are concatenated, leading to the Part-Frame-Importance (PFI) and Angle Estimator branch. The Angle Estimator module predicts view angles, enhancing the network's adaptability to view changes. An attention mechanism, the PFI module, balances the weights of distinct body parts and sequence frames. The result is an angle vector and a sequence vector, with a final classification obtained through an FC layer. Evaluations on the CASIA-B dataset showcase improved performance, with the Angle Estimator proving crucial for accuracy under different clothing conditions.

In conclusion, the aforementioned works unveil the significance of employing graph-based neural networks in the domain of gait recognition. The GCNs' ability to discern intricate relationships within skeleton sequences has paved the way for novel model-based approaches, exemplified by the Gait-D model proposed and the MAST-GCN model. These models showcase the prowess of GCN in extracting discriminative features from skeleton sequences, demonstrating robustness under challenging conditions and highlighting the adaptability of GCN in capturing nuanced patterns crucial for accurate gait identification.

5.4.3 Recurrent Neural Network

Within the gait recognition domain, Recurrent Neural Networks (RNNs) have garnered significant attention for their efficacy in capturing temporal dependencies inherent in gait sequences. An illustrative example of this approach is evident in a study employing the OpenPose algorithm to extract skeletal information [45]. This research endeavors to craft a robust gait recognition framework by focusing on extracting and fusing diverse features.

The OpenPose algorithm is crucial in delineating the skeletal structure, providing a foundational basis for subsequent feature extraction. The four primary features extracted encompass normalized 2D joint pose features, body parts, angular trajectory, and temporal displacement. These features collectively contribute to creating a comprehensive feature matrix, serving as the input for the RNN.

The RNN employed in this context incorporates a final softmax layer for streamlined classification. This architecture proves to be both straightforward and efficient, adept at discerning intricate patterns within gait sequences. Moreover, the study introduces a novel data augmentation technique involving the division of input video frames into variable-length video clips with a degree of overlap between consecutive clips. This innovative approach aims to enrich the training dataset, introducing variability and providing the model with more informative cues.

This model is evaluated on the CASIA-B dataset, showcasing commendable results in calculating recognition accuracy for each probe angle concerning various gallery angles. The evaluation extends to scenarios involving normal conditions and scenarios with bags and clothing variations. However, it is crucial to note that the evaluation metrics employed in this study may differ from conventional standards used in gait identification algorithms, potentially limiting a direct comparison with other methodologies.

This exemplar within the RNN paradigm exemplifies the amalgamation of skeletal information, feature extraction, and sequential modeling, contributing to the continuous advancements in gait recognition methodologies. Through this approach, the study showcases the potential of RNN architectures in discerning and classifying distinctive gait patterns across varying conditions and scenarios with a model-based approach.

5.4.4 Vision Transformer

This subsection discusses the employment of Vision Transformer (ViT) architectures in gait recognition, focusing on a pioneering article by Cosma et al. [7]. The study explores the application of five distinct vision transformer architectures, Simple ViT, CaiT, CrossFormer, Token2Token, and TwinsSVT, for self-supervised gait recognition.

A noteworthy aspect of this investigation lies in its adherence to a model-based approach, where the computational gait recognition system is built to understand human gait's intrinsic structure and dynamics. Each model undergoes a rigorous pre-training phase in this framework on two extensive gait datasets, GREW and DenseGait. Subsequent tests are conducted on the CASIA-B dataset, encompassing zero-shot (ZS) and fine-tuning (FT) evaluations. In the context of this study, zero-shot evaluation refers to assessing the model in conditions where it was not initially trained, such as testing on CASIA-B, a dataset distinct from the original training set. On the other hand, fine-tuning involves adapting the model on CASIA-B after the initial training, entailing a brief training period to tailor it to the specific gait recognition task.

To adapt the vision transformers for skeleton sequences, each sequence transforms into a unified skeleton image representation. The skeleton, conforming to COCO standards with 18 joints, is then spatially upsampled using bicubic interpolation to generate a square image representation suitable for most vision transformers. For instance, a skeleton sequence with 64 frames results in a corresponding skeleton image representation with dimensions of 64×18. As many vision transformers assume square images, spatial bicubic upsampling is employed to artificially increase the number of joints, constructing a square image representation for training and testing the transformer encoders. The implementation of data augmentation ensures uniform sequence length across all skeleton sequences.

Results obtained from DenseGait consistently showcase improved accuracy over GREW in the pre-training phase. Notably, the study emphasizes that SimpleViT and CrossFormer consistently achieve high performance, underscoring the discernible advantages of leveraging a pre-training dataset for downstream gait recognition tasks. The findings underscore the potential and efficacy of Vision Transformer architectures in gait recognition applications in model-based approaches, showcasing the seamless transformation of skeleton data into a square image format purposefully tailored to feed into Vision Transformers.

5.5 Deep Learning-Based Model-Free with Model-Based Approaches

Recently, a growing interest has been in exploring hybrid approaches combining model-free and model-based techniques for gait identification tasks [27]. The integration of these two paradigms offers the potential to leverage the strengths of each approach while mitigating their respective limitations. A recent study, Hua et al. [23], proposes a novel framework that combines these two approaches synergistically to enhance identification accuracy.

The framework consists of two distinct modules: a model-based module and a model-free module. The model-free module leverages deep learning techniques, particularly CNNs, to extract features from gait sequences. This module employs a two-part strategy. Firstly, the entire gait sequence is fed into a CNN to extract shallow feature information. Subsequently, global long-term features are extracted using a global long-term attention module (GLTA), and local short-term features are extracted using a designated backbone network, such as GaitPart, followed by a short-term attention module (LSTA). These features are then aggregated to form a comprehensive feature vector, further refined through training with triplet loss and cross-entropy loss.

Concurrently, the model-based module begins by extracting skeletal points from the gait sequences using the OpenPose algorithm. Subsequently, a physiological information module (HPI) is employed to extract personalized physiological features, such as shoulder width, step length, and frequency. These gait physiological features are then aggregated and fed into a fully connected neural network, also trained with triplet loss. The resulting feature vector from this module is combined with the feature vector from the model-free module to produce the final classification.

Notably, this integrated approach demonstrates effectiveness and efficiency in gait identification, achieving state-of-the-art results on benchmark datasets such as CASIA-B, particularly in challenging cross-view and varied carrying and clothing conditions. This innovative framework underscores the potential of combining model-based and model-free approaches to enhance the robustness and accuracy of gait recognition systems.

References

1. Aljazaerly, M., Makihara, Y., Muramatsu, D., Yagi, Y.: Batch hard contrastive loss and its application to cross-view gait recognition. IEEE Access **11**, 31177–31187 (2022). https://doi.org/10.1109/ACCESS.2023.3262271
2. Alsaggaf, W.A., Mehmood, I., Khairullah, E.F., Alhuraiji, S., Sabir, M., Alghamdi, A.S., Abd El-Latif, A.A.: A smart surveillance system for uncooperative gait recognition using cycle consistent generative adversarial networks (CCGANs). Comput. Intell. Neurosci. (2021). https://doi.org/10.1155/2021/3110416
3. Babaee, M., Li, L., Rigoll, G.: Person identification from partial gait cycle using fully convolutional neural networks. Neurocomputing **338**, 116–125 (2019). https://doi.org/10.1016/j.neucom.2019.01.091
4. Bekhouch, A., Bouchrika, I., Doghmane, N.: Gait biometrics: investigating the use of the lower inner regions for people identification from landmark frames. IET Biometrics **9**(6), 269–277 (2020). https://doi.org/10.1049/iet-bmt.2020.0001
5. Castro, F., Marín-Jiménez, M., Guil, N., Blanca, N.: Multimodal feature fusion for CNN-based gait recognition: an empirical comparison. Neural Comput. Appl. **32**(17), 14173–14193 (2020). https://doi.org/10.1007/s00521-020-04811-z
6. Chao, H., Wang, K., He, Y., Zhang, J., Feng, J.: GaitSet: cross-view gait recognition through utilizing gait as a deep set. IEEE Trans. Pattern Anal. Mach. Intell. **44**(7), 3467–3478 (2021). https://doi.org/10.1109/TPAMI.2021.3057879
7. Cosma, A., Catruna, A., Radoi, E.:. Exploring self-supervised vision transformers for gait recognition in the wild. Sensors **23**(5) (2023). https://doi.org/10.3390/s23052680
8. Dosovitskiy, A., Beyer, L., Kolesnikov, A., Weissenborn, D., Zhai, X., Unterthiner, T., Dehghani, M., Minderer, M., Heigold, G., Gelly, S., Uszkoreit, J., Houlsby, N.: An image is worth 16 × 16 words: transformers for image recognition at scale (2021)

9. Fan, C., Peng, Y., Cao, C., Liu, X., Hou, S., Chi, J., Huang, Y., Li, Q., He, Z.: Gaitpart: temporal part-based model for gait recognition. IEEE/CVF Conf. Comput. Vis. Pattern Recogn. (CVPR) **2020**, 14213–14221 (2020). https://doi.org/10.1109/CVPR42600.2020.01423
10. Gao, Z., Wu, J., Wu, T., Huang, R., Zhang, A., Zhao, J.: Robust clothing-independent gait recognition using hybrid part-based gait features. PeerJ Comput. Sci. **8** (2022). https://doi.org/10.7717/peerj-cs.996
11. George, M., Govindarajan, T., Rajasekaran, K., Bandi, S.: A robust similarity based deep Siamese convolutional neural network for gait recognition across views. Comput. Intell. **36**(3), 1290–1319 (2020). https://doi.org/10.1111/coin.12361
12. Ghosh, R.: A faster R-CNN and recurrent neural network based approach of gait recognition with and without carried objects. Expert Syst. Appl. **205** (2022). https://doi.org/10.1016/j.eswa.2022.117730
13. Goodfellow, I.J., Pouget-Abadie, J., Mirza, M., Xu, B., Warde-Farley, D., Ozair, S., Courville, A., Bengio, Y.: Generative adversarial networks (2014)
14. Gul, S., Malik, M.I., Khan, G.M., Shafait, F.: Multi-view gait recognition system using spatio-temporal features and deep learning. Expert Syst. Appl. **179** (2021). https://doi.org/10.1016/j.eswa.2021.115057
15. Guo, H., Li, B., Zhang, Y., Zhang, Y., Li, W., Qiao, F., Rong, X., Zhou, S.: Gait recognition based on the feature extraction of gabor filter and linear discriminant analysis and improved local coupled extreme learning machine. Math. Prob. Eng. **2020** (2020). https://doi.org/10.1155/2020/5393058
16. Gupta, S., Chattopadhyay, P.: Exploiting pose dynamics for human recognition from their gait signatures. Multimedia Tools Appl. **80**(28–29), 35903–35921 (2020). https://doi.org/10.1007/s11042-020-10071-9
17. Gupta, S., Chattopadhyay, P.: Gait recognition in the presence of co-variate conditions. Neurocomputing **454**, 76–87 (2021). https://doi.org/10.1016/j.neucom.2021.04.113
18. Gupta, S., Chattopadhyay, P.: Pose-based boundary energy image for gait recognition from silhouette contours. Sadhana—Acad. Proc. Eng. Sci. **48**(4) (2023). https://doi.org/10.1007/s12046-023-02234-9
19. Han, F., Li, X., Zhao, J., Shen, F.: A unified perspective of classification-based loss and distance-based loss for cross-view gait recognition. Pattern Recogn. **125** (2022). https://doi.org/10.1016/j.patcog.2021.108519
20. Hawas, A., El-Khobby, H., Elnaby, M., El-Samie, F.: Gait identification by convolutional neural networks and optical flow. Multimedia Tools Appl. **78**(18), 25873–25888 (2019). https://doi.org/10.1007/s11042-019-7638-9
21. He, K., Zhang, X., Ren, S., Sun, J.: Deep residual learning for image recognition (2015)
22. Hou, S., Liu, X., Cao, C., Huang, Y.: Gait quality aware network: toward the interpretability of silhouette-based gait recognition. IEEE Trans. Neural Netw. Learn. Syst. **34**(11), 8978–8988 (2022). https://doi.org/10.1109/TNNLS.2022.3154723
23. Hua, C., Pan, Y., Li, J., Wang, Z.: Gait recognition by combining the long-short-term attention network and personal physiological features. Sensors **22**(22) (2022). https://doi.org/10.3390/s22228779
24. Iwashita, Y., Sakano, H., Kurazume, R., Stoica, A.: Speed invariant gait recognition—the enhanced mutual subspace method. PLoS ONE **16**(8 August) (2021). https://doi.org/10.1371/journal.pone.0255927
25. Jameel, H., Dhannoon, B.: Gait recognition based on deep learning. Iraqi J. Sci. **63**(1), 397–408 (2021). https://doi.org/10.24996/ijs.2022.63.1.36
26. Jia, P., Zhao, Q., Li, B., Zhang, J.: CJAM: convolutional neural network joint attention mechanism in gait recognition. IEICE Trans. Inform. Syst. **E104D**(8), 1239–1249 (2021). https://doi.org/10.1587/transinf.2020BDP0010
27. Khaliluzzaman, M., Uddin, A., Deb, K., Hasan, M.J.: Person recognition based on deep gait: a survey. Sensors, **23**(10) (2023). https://doi.org/10.3390/s23104875
28. Li, X., Makihara, Y., Xu, C., Yagi, Y., Ren, M.: Joint intensity transformer network for gait recognition robust against clothing and carrying status. IEEE Trans. Inform. Forensics Secur. **14**(12), 3102–3115 (2019). https://doi.org/10.1109/TIFS.2019.2912577

29. Lima, V., Melo, V., Schwartz, W.: Simple and efficient pose-based gait recognition method for challenging environments. Pattern Anal Appl **24**(2), 497–507 (2020). https://doi.org/10.1007/s10044-020-00935-z
30. Lin, B., Zhang, S., Wang, M., Li, L., Yu, X.: GaitGL: learning discriminative global-local feature representations for gait recognition (2022)
31. Linda, G.M., Themozhi, G., Bandi, S.R.: Color-mapped contour gait image for cross-view gait recognition using deep convolutional neural network. Int. J. Wavelets, Multiresolut. Inform. Process. **18**(1) (2019). https://doi.org/10.1142/S0219691319410121
32. Marin-Jimenez, M.J., Castro, F.M., Delgado-Escano, R., Kalogeiton, V., Guil, N.: UGaitNet: multimodal gait recognition with missing input modalities. IEEE Trans. Inform. Forensics Secur. **16**, 5452–5462 (2021). https://doi.org/10.1109/TIFS.2021.3132579
33. Merlin Linda, G., Lakshmi, Sree Rathna, N.V.S., Murugan, N.S., Mahapatra, R.P., Muthukumaran, V., Sivaram, M.: Intelligent recognition system for viewpoint variations on gait and speech using CNN-CapsNet. Int. J. Intell. Computi. Cybern. **15**(3):363–382 (2021). https://doi.org/10.1108/IJICC-08-2021-0178
34. Mogan, J., Lee, C., Lim, K., Ali, M., Alqahtani, A.: Gait-CNN-ViT: multi-model gait recognition with convolutional neural networks and vision transformer. Sensors **23**(8) (2023). https://doi.org/10.3390/s23083809
35. Mogan, J., Lee, C., Lim, K., Muthu, K.: VGG16-MLP: gait recognition with fine-tuned VGG-16 and multilayer perceptron. Appl. Sci. (Switzerl.) **12**(15) (2022). https://doi.org/10.3390/app12157639
36. Mogan, J.N., Lee, C.P., Lim, K.M., Muthu, K.S.: Gait-ViT: gait recognition with vision transformer. Sensors **22**(19) (2022). https://doi.org/10.3390/s22197362
37. Moghaddam, A., Etemad, A.: View-invariant gait recognition with attentive recurrent learning of partial representations. IEEE Trans. Biometrics Behav. Identity Sci. **3**(1), 124–137 (2020). https://doi.org/10.1109/TBIOM.2020.3031470
38. Pinčić, D., Sušanj, D., Lenac, K.: Gait recognition with self-supervised learning of gait features based on vision transformers. Sensors **22**(19) (2022). https://doi.org/10.3390/s22197140
39. Sabour, S., Frosst, N., Hinton, G.E.: Dynamic routing between capsules. Adv. Neural Inform. Process. Syst. **30** (2017)
40. Sayeed, M.S., Min, P.P., Bari, M.A.: Deep learning based gait recognition using convolutional neural network in the COVID-19 pandemic. Emerg. Sci. J. **6**(5), 1086–1099 (2022). https://doi.org/10.28991/ESJ-2022-06-05-012
41. Sepas-Moghaddam, A., Ghorbani, S., Troje, N.F., Etemad, A.: Gait recognition using multi-scale partial representation transformation with capsules (2020)
42. Sharma, P., Khan, A., Jain, S., Srivastava, A.: Artificial intelligence driven human identification. J. Inform. Technol. Manag. **15**(3), 113–133 (2023). https://doi.org/10.22059/JITM.2023.93629
43. Shiraga, K., Makihara, Y., Muramatsu, D., Echigo, T., Yagi, Y.: Geinet: view-invariant gait recognition using a convolutional neural network. Int. Conf. Biometrics (ICB) **2016**, 1–8 (2016). https://doi.org/10.1109/ICB.2016.7550060
44. Thapar, D., Jaswal, G., Nigam, A., Arora, C.: Gait metric learning Siamese network exploiting dual of spatio-temporal 3D-CNN intra and LSTM based inter gait-cycle-segment features. Pattern Recogn. Lett. **125**, 646–653 (2019). https://doi.org/10.1016/j.patrec.2019.07.008
45. Upadhyay, J., Gonsalves, T.: An enhanced gait recognition system based on the features fusion methodology with recurrent neural network (RNN). Indian J. Comput. Sci. Eng. **13**(5), 1483–1496 (2022). https://doi.org/10.21817/indjcse/2022/v13i5/221305042
46. Wang, R., Shi, Y., Ling, H., Li, Z., Li, P., Liu, B., Zheng, H., Wang, Q.: Gait recognition via gait period set. IEEE Trans. Biometrics Behav. Identity Sci. **5**(2), 183–195 (2023). https://doi.org/10.1109/TBIOM.2023.3244206
47. Wang, X., Hu, S.: Visual gait recognition based on convolutional block attention network. Multimedia Tools Appl. **81**(20), 29459–29476 (2022). https://doi.org/10.1007/s11042-022-12831-1
48. Wang, X., Yan, W.Q.: Human gait recognition based on frame-by-frame gait energy images and convolutional long short-term memory. Int. J. Neural Syst. **30**(1) (2019). https://doi.org/10.1142/S0129065719500278

49. Wang, X., Zhang, J.: Gait feature extraction and gait classification using two-branch CNN. Multimedia Tools Appl. **79**(3–4), 2917–2930 (2019). https://doi.org/10.1007/s11042-019-08509-w
50. Wang, Y., Chen, Z., Wu, Q., Rong, X.: Deep mutual learning network for gait recognition. Multimedia Tools Appl. **79**(31–32), 22653–22672 (2020). https://doi.org/10.1007/s11042-020-09003-4
51. Wang, Y., Song, C., Huang, Y., Wang, Z., Wang, L.: Learning view invariant gait features with Two-Stream GAN. Neurocomputing **339**, 245–254 (2019). https://doi.org/10.1016/j.neucom.2019.02.025
52. Wen, J., Wang, X.: Cross-view gait recognition based on residual long short-term memory. Multimedia Tools Appl. **80**(19), 28777–28788 (2021). https://doi.org/10.1007/s11042-021-11107-4
53. Wen, J., Wang, X.: Gait recognition based on sparse linear subspace. IET Image Process. **15**(12), 2761–2769 (2021). https://doi.org/10.1049/ipr2.12260
54. Wu, C., Song, Y., Zhang, Y.: Multi-view gait recognition using NMF and 2DLDA. Multimedia Tools Appl. **78**(24), 35789–35811 (2019). https://doi.org/10.1007/s11042-019-08153-4
55. Wu, H., Tian, J., Fu, Y., Li, B., Li, X.: Condition-aware comparison scheme for gait recognition. IEEE Trans. Image Process. **30**, 2734–2744 (2020). https://doi.org/10.1109/TIP.2020.3039888
56. Xu, K., Jiang, X., Sun, T.: Gait recognition based on local graphical skeleton descriptor with pairwise similarity network. IEEE Trans. Multimedia **24**, 3265–3275 (2021). https://doi.org/10.1109/TMM.2021.3095809
57. Xu, Z., Lu, W., Zhang, Q., Yeung, Y., Chen, X.: Gait recognition based on capsule network. J. Vis. Commun. Image Represent. **59**, 159–167 (2019). https://doi.org/10.1016/j.jvcir.2019.01.023
58. Xue, W., Ai, H., Sun, T., Song, C., Huang, Y., Wang, L.: Frame-GAN: increasing the frame rate of gait videos with generative adversarial networks. Neurocomputing **380**, 95–104 (2019). https://doi.org/10.1016/j.neucom.2019.11.015
59. Yao, L., Kusakunniran, W., Zhang, P., Wu, Q., Zhang, J.: Improving disentangled representation learning for gait recognition using group supervision. IEEE Trans. Multimedia (2022). https://doi.org/10.1109/TMM.2022.3171961
60. Yeo, S.-S., Rho, S., Kim, H., Safdar, J., Zia, U., Durrani, M.Y.: A triplet-branch convolutional neural network for part-based gait recognition. Comput. Syst. Sci. Eng. **47**(2), 2027–2047 (2023). https://doi.org/10.32604/csse.2023.040327
61. Yu, Y., Si, X., Hu, C., Zhang, J.: A review of recurrent neural networks: LSTM cells and network architectures. Neural Comput. **31**(7), 1235–1270 (2019). https://doi.org/10.1162/neco_a_01199
62. Zhang, Y., Huang, Y., Yu, S., Wang, L.: Cross-view gait recognition by discriminative feature learning. IEEE Trans. Image Process. **29**, 1001–1015 (2019). https://doi.org/10.1109/TIP.2019.2926208
63. Zheng, L., Zha, Y., Kong, D., Yang, H., Zhang, Y.: Multi-branch angle aware spatial temporal graph convolutional neural network for model-based gait recognition. IET Cyber-Syst. Robot. **4**(2), 97–106 (2022). https://doi.org/10.1049/csy2.12052
64. Zhou, J., Cui, G., Hu, S., Zhang, Z., Yang, C., Liu, Z., Wang, L., Li, C., Sun, M.: Graph neural networks: a review of methods and applications. AI Open **1**, 57–81 (2020). https://doi.org/10.1016/j.aiopen.2021.01.001

Chapter 6
Conclusion

Abstract This conclusion highlights key insights from the study presented, emphasizing the evolving landscape of gait recognition from traditional machine learning to deep learning-based methods. While identifying ongoing challenges such as data quality, ethical considerations, and architectural limitations, this chapter points towards the promising convergence of model-free and model-based approaches. This amalgamation is poised to drive future innovations, enhancing the accuracy and robustness of gait recognition systems for diverse real-world applications. It concludes by outlining potential avenues for future research and development in the field.

6.1 Tying It All Together: Key Insights from the Study

Traditional machine learning-based gait identification methods share a reliance on established techniques such as GEIs, PCA, and distance metrics. Their strengths lie in limited-data scenarios, where careful feature engineering and dimensionality reduction contribute to robust performance. However, the increasing prominence of deep learning-based approaches, especially in handling larger datasets, suggests a shifting landscape in gait identification research, with more emphasis on automatic feature learning and end-to-end models.

The gait recognition field has recently gained substantial attention due to its diverse applications that can be found in surveillance, healthcare, and biometric identification. Despite advancements, challenges persist, such as individual variations, clothing impact, carrying conditions, and walking speed variations. Input data quality, including resolution, illumination, and occlusion, significantly affects system performance. Ethical considerations, like privacy concerns and misidentification risks, further complicate gait recognition. Robust systems overcoming these challenges are essential for real-world applications. The covariate factors affecting accuracy include carrying conditions, clothing variations, viewpoint differences, occlusion, cross-view conditions, speed variations, unconstrained environments, and spatial-temporal complexities. Deep architecture for gait recognition encounters challenges,

D. R. M. Bastos and J. M. R. S. Tavares, *Advances in Gait-Based Identification*, Studies in Systems, Decision and Control 593,
https://doi.org/10.1007/978-3-031-89560-9_6

including limited labeled data availability, designing effective architectures capturing spatial and temporal information, and ensuring robustness to environmental variations. Interpretability also remains a challenge.

Model-free approaches focusing on silhouette encounter limitations with covariate factors. Model-based methods analyzing skeleton data face challenges in extracting accurate data, interdependency on pose estimation, spatial-temporal complexities, and viewpoint sensitivity. Overcoming silhouette image problems, skeleton structures offer advantages in gait recognition. However, deep neural architectures face issues in treating skeleton data, such as limited representation, rotation invariance, difficulty handling different angles, and struggles to reveal spatial connections. CapsNets and GNNs address some of these challenges, allowing for more complex relationship capture in gait data.

6.2 Perspectives and Future Directions in Gait Recognition

The challenges underscore the need for continued research and development to enhance gait recognition technology and allow its application in a real-world context.

In the foreseeable future, the convergence of model-free and model-based approaches is expected to become increasingly prevalent in the field of gait identification. Their integration is anticipated to emerge as a standard practice for achieving superior performance and robustness in gait recognition systems. By combining the data-driven capabilities of model-free methods with the structural constraints offered by model-based approaches, future systems can leverage the best of both worlds to overcome each approach's inherent limitations. This fusion promises to unlock new avenues for innovation, allowing more effective use of available data and domain knowledge to yield more accurate, reliable, and adaptable gait identification solutions.

The landscape of model-free and model-based approaches in gait recognition is dynamic and continually evolving. As researchers explore novel architectures, loss functions, and attention mechanisms, the field holds promise for even more significant advancements in the accuracy, efficiency, and real-world applicability of gait recognition systems. The findings from these studies collectively contribute to shaping the future trajectory of gait recognition research, guiding the development of more resilient and sophisticated identification techniques.

The manufacturer's authorised representative in the EU is Springer Nature Customer Service Centre GmbH, Europaplatz 3, 69115 Heidelberg, Germany. If you have any concerns regarding our products, please contact ProductSafety@springernature.com

Printed and bound by CPI Group (UK) Ltd, Croydon, CR0 4YY
01/10/2026
02225875-0004